Atlas of Structural Geology

Atlas of Structural Geology

Soumyajit Mukherjee
Department of Earth Sciences
Indian Institute of Technology Bombay
Powai, Mumbai 400 076
Maharashtra, India
soumyajitm@gmail.com

ELSEVIER

AMSTERDAM • BOSTON • HEIDELBERG • LONDON • NEW YORK • OXFORD • PARIS
SAN DIEGO • SAN FRANCISCO • SINGAPORE • SYDNEY • TOKYO

Elsevier
Radarweg 29, PO Box 211, 1000 AE Amsterdam, Netherlands
The Boulevard, Langford Lane, Kidlington, Oxford OX5 1GB, UK
225 Wyman Street, Waltham, MA 02451, USA

ISBN: 978-0-12-420152-1

British Library Cataloguing in Publication Data
A catalogue record for this book is available from the British Library

Library of Congress Cataloging-in-Publication Data
A catalog record for this book is available from the Library of Congress

For Information on all Elsevier publications
visit our website at http://store.elsevier.com/

Printed and bound in the USA

Contents

List of Contributors vii
Preface xiii
Acknowledgments xv

1. Folds 1
2. Ductile Shear Zones 49
3. Brittle Faults 79
4. Boudins and Mullions 107
5. Veins 119
6. Various Structures 125

Author Index 159
Subject Index 163

List of Contributors

Max Arndt, E-mail: max.arndt@emr.rwth-aachen.de Info: EMR - Energy & Mineral Resources Group, Geologie - Endogene Dynamik, RWTH Aachen University, Lochnerstrasse 4-20, D-52056 Aachen, Germany; Tel: +49-241-80-98438

Paola Ferreira Barbosa, E-mail: paolafeba@yahoo.com.br Info: Center of Microscopy, Universidade Federal de MInas Gerais, Avenida Perimetral Sul, 91-129 - Campus da UFMG, Belo Horizonte - MG, Brazil; CEP: 31255-040; Tel: (55 31) 3409-7575

Ananya Basu, E-mail: an.basuju@gmail.com Info: Department of Geological Sciences, Jadavpur University, Kolkata, West Bengal, India; Pin-700 032; 0919007684783 (M)

Andrea Billi, E-mail: andrea.billi@cnr.it Info: Consiglio Nazionale delle Ricerche, IGAG, c.o. Dipartimento Scienze della Terra, Sapienza Università di Roma, P.le A. Moro 5, 00185, Rome, Italy; 390649914955 (M)

Tuhin Biswas, E-mail: tbtuhin24@gmail.com Info: Department of Earth Sciences, Indian Institute of Technology Bombay, Powai, Mumbai, Maharashtra, India; Pin-400 076; 919167517063 (M)

Chloë Bonamici, E-mail: chloebee@lanl.gov Info: Los Alamos National Laboratory, Chemistry Division (C-NR), Los Alamos, NM 87545; +1 505 664 0115

Svetoslav Bontchev, E-mail: srbontchev@gmail.com Info: Graf Ignatiev St. 53 G 1142, Sofia, Bulgaria; 359-884 338 115

Narayan Bose, E-mail: narayan.bghs@gmail.com Info: Department of Earth Sciences, Indian Institute of Technology Bombay, Powai, Mumbai, Maharashtra, India; Pin-400 076; 919029787238 (M)

Luis A. Buatois, E-mail: luis.buatois@usask.ca Info: Department of Geological Sciences, University of Saskatchewan, 114 Science Place, Saskatoon, Saskatchewan, Canada, S7N5E2; Phone: 1-306-966-5730

Paul K. Byrne, E-mail: pbyrne@carnegiescience.edu Info: Lunar and Planetary Institute, Universities Space Research Association, 3600 Bay Area Blvd., Houston, TX 77058; Phone office: 281-486-2140

Jonathan Saul Caine, E-mail: jscaine@usgs.gov Info: U. S. Geological Survey, P. O. Box 25046, MS 964, Denver, CO 80225-0046; 303 236 1822

Sadhana M. Chatterjee, E-mail: sadhanamahato@gmail.com Info: Department of Geological Sciences, Jadavpur University, Kolkata, West Bengal, India; Pin-700 032; 9434217749 (M)

Sreejita Chatterjee, E-mail: sreejitach@gmail.com Info: Department Earth Sciences, Indian Institute of Technology Bombay, Powai, Mumbai, Maharashtra, India; Pin-400 076; 8879048571 (M)

T.R.K. Chetty, E-mail: trkchetty@gmail.com Info: National Geophysical Research Institute, Hyderabad, Andhra Pradesh, India; Pin-500 007; 9885676086 (M)

Mainak Choudhuri, E-mail: mainak_ch@yahoo.co.in Info: Reliance Industries Limited, Petroleum Business (E & P), Mumbai, India; Pin-400 071; 919967535923 (M)

Sankha Das, E-mail: sankhad56@gmail.com Info: Geological Survey of India, State Unit, Andhra Pradesh, Southern Region, Bandlaguda, Hyderabad, Andhra Pradesh; Pin-500068; 09642276706 (M), 09830154951 (M)

Rohini Das, E-mail: romiyadas@gmail.com Info: Department of Earth Sciences, Indian Institute of Technology Roorkee, Roorkee, Uttarakhand, India; Pin-247 667; 8126694552 (M), 9831518422 (M)

Sudipta Dasgupta, E-mail: sudipta.dasgupta@usask.ca Info: Department of Geological Sciences, University of Saskatchewan, 114 Science Place, Saskatoon, Saskatchewan, Canada S7N 5E2; 13069662457 (O)

Swagato Dasgupta, E-mail: swagato.dg@gmail.com Info: Reliance Industries Ltd., Exploration & Production, Navi Mumbai, Maharashtra, India; Pin-400 701; 919987538641 (M)

Bhushan S. Deota, E-mail: bdeota@rediffmail.com Info: Department of Geology, Faculty of Science, The Maharaja Sayajirao University of Baroda, Vadodara, Gujarat; 390002; 9898097211 (M)

Tine Derez, E-mail: Tine.Derez@ees.kuleuven.be Info: Department of Earth and Environmental Sciences, KU Leuven, Celestijnenlaan 200E box 2410, B-3001 Heverlee, Belgium; +32(0)496502579

Natalie Deseta, E-mail: suridae@gmail.com Info: School of Geosciences, University of the Witwatersrand, Private Bag 3, WITS 2050; 27796096295 (M)

Arindam Dutta, E-mail: arindamdutta2000@gmail.com Info: Department of Geological Sciences, Indian Institute of Technology Kharagpur, India; 9830956422 (M)

Dripta Dutta, E-mail: dripta.dutta@gmail.com Info: Department of Earth Sciences, Indian Institute of Technology Bombay, Powai, Mumbai, Maharashtra, India; Pin-400 076; 919167521824 (M)

Amy Ellis, E-mail: amyellis@hotmail.co.uk Info: Ikon Science Ltd, Rivergreen Centre, Durham, DH1 5TS, UK; +44 (0) 191 383 7362 (M)

Ake Fagereng, E-mail: FagerengA@cardiff.ac.uk Info: School of Earth & Ocean Science, Cardiff University, Main Building, Park Place, Cardiff, CF10 3AT, UK; +44 (0)29 208 70760 (O)

Carlos Fernández, E-mail: fcarlos@uhu.es Info: Departamento Geodinámica y Paleontologia, Facultad de Ciencias Experimentales, Universidad de Huelva, Campus El Carmen, Avenida 3 de Marzo, 21071 Huelva, Spain; Phone: +34 959219857

Luigi De Filippis, E-mail: luigidefilippis1@gmail.com Info: Dipartimento di Scienze, Università Roma Tre, Largo S.L. Murialdo 1, I-00146 Roma; +39-3286138869

László Fodor, E-mail: asz.fodor@yahoo.com Info: Hungarian Academy of Sciences MTA-ELTE, Geological, Geophysical and Space Sciences, Research Group at Eötvös University, Pázmány P. sétány 1/C; Phone: 36-1-3722500/8714

Chiara Frassi, E-mail: chiarafrassi@yahoo.it Info: Dipartimento di Scienze della Terra, Università di Pisa, via S. Maria, 53, 56100 Pisa, Italy; +39 050 2215781

M.S. Gadhavi, E-mail: mahendrasinh@gmail.com Info: Civil Engineering Department, L. D. College of Engineering, Ahmedabad, 380 015 Gujarat, India; +91-9426272328 (M)

Rajkumar Ghosh, E-mail: rajkumarghgeol@gmail.com Info: Department of Earth Sciences, Indian Institute of Technology Bombay, Powai, Mumbai, Maharashtra, India; Pin-400 076; 9167520347 (M)

Guido Gosso, E-mail: Guido.Gosso@unimi.it Info: Department of Earth Sciences, Universita' degli Studi di Milano; Citta' Studi, Via Mangiagalli 34, 20133 Milano, Italy; 3902 5031 55 55

Tapos Kumar Goswami, E-mail: taposgoswami@gmail.com Info: Department of Applied Geology, Dibrugarh University, Dibrugarh, Assam, India; Pin-786 004; 919435352889 (M)

Sukanta Goswami, E-mail: sukantagoswami@iitb.ac.in Info: Atomic Minerals Directorate for Exploration and Research (AMD), Department of Atomic Energy (DAE), Southern Region, Nagarbhavi, Bangalore, India; Pin-560 072; 8088872383 (M)

Jens Carsten Grimmer, E-mail: jens.grimmer@kit.edu Info: Institute of Applied Geosciences, Adenauerring 20b, 76131 Karlsruhe, Germany; +49 721 608 41888

Ranjan Gupta, E-mail: gupta.ranjan256@gmail.com Info: Rampura Agucha Mine, Hindustan Zinc Limited, Rajasthan, India; 91-9799175999 (M)

Saibal Gupta, E-mail: saibl2008@gmail.com Info: Department of Geology and Geophysics, Indian Institute of Technology Kharagpur, Kharagpur, West Bengal, India; Pin: 721302; +91-3222-283370 (M)

Tomokazu Hokada, E-mail: hokada@nipr.ac.jp Info: Geology Group, National Institute of Polar Research, 10-3 Midori-cho, Tachikawa, Tokyo 190-8518, Japan; Phone: +81-42-512-0714

Guillermo Alvarado Induni, E-mail: GAlvaradoI@ice.go.cr Info: Área de Amenazas y Auscultación Sismológica y Volcánica, C.S. Exploración Subterránea/ NIC-Electricidad. Apdo 10032-1000; Landline: 00506 89383752

Scott Johnson, E-mail: johnsons@maine.edu Info: School of Earth and Climate Sciences, 5790 Bryand Global Sciences, University of Maine, Orono, ME 04469-5790, USA; (207)581-2142

Aditya Joshi, E-mail: adityaujoshi@gmail.com Info: Department of Geology, Faculty of Science, The M.S. University of Baroda, Vadodara, Gujarat, India; Pin-390 002; 918000421451 (M)

Eirin Kar, E-mail: sakurakar5@gmail.com Info: Department of Earth Sciences, Indian Institute of Technology Roorkee, Roorkee, Uttarakhand, India; Pin-247 667; 8439731513 (M)

Rahul Kar, E-mail: rkar48@gmail.com Info: Department of Geology and Geophysics, Indian Institute of Technology Kharagpur, West Bengal, India; Pin-721 302; 919749469334 (M)

R.V. Karanth, E-mail: r_v_karanth@yahoo.co.in Info: 104 - Aarth Apartments, 29 - Pratapgunj, Vadodara - 390 002, Gujarat, India; 919998485468 (M)

Miklós Kázmér, E-mail: mkazmer@gmail.com Info: Department of Palaeontology, Eotvos University, Pazmany Peter setany 1/c; 36-20-494-5275

Subodha Khanal, E-mail: skhanal@crimson.ua.edu Info: Department of Geological Sciences, University of Alabama, Tuscaloosa, Al, 35487, USA; 1(347)400-3645

Christian Klimczak, E-mail: cklimczak@ciw.edu Info: Department of Geology, 210 Field Street, University of Georgia, Athens, GA 30602-2501; 706-542-2977

Leonardo Evangelista Lagoeiro, E-mail: lagoeiro@icloud.com Info: Microanalysis and Microscopy Laboratory – MICROLAB, Universidade Federal de Ouro Preto, Geology Department, Morro do Cruzeiro, s/n, Ouro Preto, MG, CEP: 354000-000, Minas Gerais MG, Brazil; Tel: (55 31) 3559-1859; (55 31) 9838-5328 (M)

Mariano A. Larrovere, E-mail: marianlarro@gmail.com Info: Centro Regional de Investigaciones Científicas y Transferencia Tecnológica de La Rioja (CRILAR-CONICET). Entre Ríos y Mendoza s/n. Anillaco (5301), La Rioja, Argentina; INGeReN-CENIIT-UNLaR. Av. Gob. Vernet y Apóstol Felipe, La Rioja (5300), Argentina; 540387715440689 (M)

M.A. Limaye, E-mail: manoj_geol@rediffmail.com Department of Geology, Faculty of Science, The M.S.University of Baroda, Vadodara, Gujarat, India; Phone: 91 9824071285 (M)

Esther Izquierdo-Llavall, E-mail: estheriz@unizar.es Info: Departamento Ciencias de la Tierra, Universidad de Zaragoza, Pedro Cerbuna St., 12, 50.009 Zaragoza, Spain; Phone: +34-976-762127

Shengli Ma, E-mail: masl@ies.ac.cn Info: State Key Laboratory of Earthquake Dynamics, Institute of Geology, China Earthquake Administration, P. O. Box 9803, Beijing 100029, China; Tel: 86-13426054959

Kankajit Maji, E-mail: kankajit.maji0@gmail.com Info: Department of Earth Sciences, Indian Institute of Technology Bombay, Powai, Mumbai, Maharashtra, India; Pin-400 076; +91-022-2576-7281

Neil Mancktelow, E-mail: neil.mancktelow@erdw.ethz.ch Info: Geological Institute, ETH Zurich, CH-8092 Zurich, Switzerland; +41 44 632 3671

Subhadip Mandal, E-mail: smandal@crimson.ua.edu Info: Department of Geological Sciences, University of Alabama, 201 7th Avenue, Room No. 2003, Bevill Building, Tuscaloosa, AL 35487, USA; +1 (205) 348-5095

George Mathew, E-mail: gmathew@iitb.ac.in Info: Department of Earth Sciences, Indian Institute of Technology Bombay, Powai, Mumbai, Maharashtra, India; Pin-400 076; 919820287275 (M)

Francesco Mazzarini, E-mail: mazzarini@pi.ingv.it Info: Istituto Nazionale di Geofisica e Vulcanologia, Sezione di Pisa, Via della Faggiola 32, 56126 Pisa, Italy; Tel: +39 050 8311956

Patrick Meere, E-mail: p.meere@ucc.ie Info: School of Biological, Earth & Environmental Sciences, University College Cork, Cork, Ireland; T: +353-21-490-3000; F: +353 21 490 3000

Achyuta Ayan Misra, E-mail: achyutaayan@gmail.com Info: Department of Earth Sciences, Indian Institute of Technology Bombay, Powai, Mumbai, Maharashtra, India; Pin-400 076; Reliance Industries Limited, Petroleum Business (E & P), Mumbai, India; Pin-400 071; 919967017133 (M)

Atanu Mukherjee, E-mail: atanu.pathor@gmail.com Info: Atomic Minerals Directorate for Exploration and Research, Department of Atomic Energy, Bangalore, India; Pin-500 016; 7483028746 (M)

Soumyajit Mukherjee, E-mail: soumyajitm@gmail.com Info: Department of Earth Sciences, Indian Institute of Technology Bombay, Powai, Mumbai, Maharashtra, India; Pin-400 076; 9167625339 (M)

Kieran F. Mulchrone, E-mail: k.mulchrone@ucc.ie Info: Department of Applied Mathematics, School of Mathematical Sciences, University College, Cork, Ireland; 353214205822 (O)

Giovanni Musumeci, E-mail: gm@unipi.it Info: Dipartimento di Scienze della Terra, Università di Pisa, Via S. Maria, 53, 56126 Pisa, Italy; Tel: +39 050 2215745

Shruthi Narayanan, E-mail: 123060007@iitb.ac.in Info: Department of Earth Sciences, Indian Institute of Technology Bombay, Powai, Mumbai, Mahatashtra, India; Pin-400 076; 91-9967042788 (M)

Payman Navabpour, E-mail: payman.navabpour@gmail.com Info: Friedrich Schiller Universität Jena, Institut für Geowissenschaften, Jena, Germany; 49-17672244672

Lucie Novakova, E-mail: lucie.novakova@irsm.cas.cz Info: Department of Seismotectonics, Institute of Rock Structure and Mechanics, Academy of Sciences of the Czech Republic, V Holešovičkách 41, 182 09 Prague 8; Phone: +420605117392

Belén Oliva-Urcia, E-mail: boliva@unizar.es Info: Departamento de Ciencias de la Tierra, Universidad de Zaragoza, 50009 Zaragoza, Spain; +34 976 762 127 (M)

Yasuhito Osanai, E-mail: osanai@scs.kyushu-u.ac.jp Info: Division of Earth Sciences, Department of Environmental Changes, Faculty of Social and Cultural Studies, Kyushu University, 744 Motooka, Fukuoka, 819-0395 Japan; Phone: +81-92-802-5660

Masaaki Owada, E-mail: owada@sci.yamaguchi-u.ac.jp Info: Graduate school of Science and Technology, Division of Earth Sciences, Yamaguchi University, Yoshida 1677-1, Yamaguchi, 753-8512 Japan; Phone: +81-83-933-5751

Paolo Pace, E-mail: p.pace@unich.it Info: Dipartimento di Ingegneria e Geologia, Università degli Studi "G. d'Annunzio" di Chieti Pescara, Via dei Vestini, 31, 66013, Chieti Scalo (CH), Italy; 393490825427 (M)

Jorge Pamplona, E-mail: jopamp@dct.uminho.pt Info: ICT, Departamento de Ciências da Terra, Escola de Ciências, Universidade do Minho, Campus de Gualtar, 4710-057 Braga, Portugal; +351 253604300

M.K. Panigrahi, E-mail: mkp@gg.iitkgp.ernet.in Info: Department of Geology and Geophysics, Indian Institute of Technology Kharagpur, Kharagpur, West Bengal, India; Pin: 721302; +91-3222-283376 (O)

Jyotirmoy Paul, E-mail: djyo.geos01@gmail.com Info: Department of Geological Sciences, Jadavpur University, Kolkata, West Bengal, India; Pin-700 032; 0919051469485 (M)

Victoria Pease, E-mail: vicky.pease@geo.su.se Info: Department of Geological Sciences, PetroTectonics Facility, Stockholm University, SE-106 91 Stockholm, Sweden; 468674-7321

Giorgio Pennacchioni, E-mail: giorgio.pennacchioni@unipd.it Info: Dipartimento di Geoscienze, University of Padova, Via Gradenigo 6, Italy; +39 338 6718488 (M)

Roberto Vizeu Lima Pinheiro, E-mail: vizeu@ufpa.br Info: Universidade Federal do Para - Faculdade de Geologia - Brazil; (0055 91) 32017393

Suellen Olívia Cândida Pinto, E-mail: suellen_olivia@yahoo.com.br Info: Federal University of OuroPreto, Geology Engineering Department, Ouro Preto, Brazil

Andrés Pocoví, E-mail: apocovi@unizar.es Info: Geotransfer Res. Group, Department of Earth Sciences, Pedro Cerbuna 12, 50009 Zaragoza; +34 976 76 20 72

Brian R. Pratt, E-mail: brian.pratt@usask.ca Info: Department of Geological Sciences, University of Saskatchewan, 114 Science Place, Saskatoon, Saskatchewan, Canada, S7N5E2; Phone: 1-306-966-5725

Emilio L. Pueyo, E-mail: unaim@igme.es Info:Unidad de Zaragoza, Instituto Geologico y Minero de España, C/ Manuel Lasala 44, 9°B, 50006 Zaragoza, Spain; +34 976 55 51 53 (ext 31)

Benedito Calejo Rodrigues, E-mail: bjcrodrigues@gmail.com Info: ICT, Rua do Campo Alegre 687, 4169-007 Porto, Portugal; +351 220402472

Federico Rossetti, E-mail: federico.rossetti@uniroma3.it Info: Dipartimento di Scienze, Università degli Studi Roma Tre, L.go S.L. Murialdo, 1, 00146 Rome, Italy; +39 06 57338043 (M)

Rajib Sadhu, E-mail: imrajib.geo@gmail.com Info: Office address: Premiere Miniere Du Katanga P.M.K. Sprl 4 Avenue Des Cypres N R C: 10316 ID. NAT:6-118-N60935L Lubumbashi R.D.Congo; 243-976325555

Dilip Saha, E-mail: sahad.geol@gmail.com Info: Geological Studies Unit, Indian Statistical Institute, 203 B T Road, Kolkata, India; Pin-700 108; 919433559563 (M)

Dnyanada Salvi, E-mail: salvidnyanada@gmail.com Info: Department of Earth Sciences, Indian Institute of Technology Bombay Powai, Mumbai 400 076, Maharashtra, India; Phone: +91-022-2576-7251; 9869007336 (M)

Anupam Samanta, E-mail: anupam.jugeology@gmail.com Info: Department of Geological Sciences, Jadavpur University, Kolkata, West Bengal, India; Pin-700 032; 91-9046368545 (M)

Elisa M. Sánchez, E-mail: emsanchez@ubu.es Info: Paleomagnetic Laboratory, Physics Dept. Universidad de Burgos, Avda. de Cantabria, s/n, 09006 Burgos, Spain; +34 947 25 8978

Moloy Sarkar, E-mail: moloy.sarkar1992@gmail.com Info: Department of Earth Sciences, Indian Institute of Technology Roorkee, Roorkee, Uttarakhand, India; Pin-247 667; 09038850981 (M), 08791258153 (M)

Jennifer J. Scott, E-mail: jjscott@ualberta.ca Info: Department of Geological Sciences, University of Saskatchewan, 114 Science Place, Saskatoon, SK, Canada S7N 5E2

Souvik Sen, E-mail: souvikseniitb@gmail.com Info: Department of Earth Sciences, Indian Institute of Technology Bombay, Powai, Mumbai, Mahatashtra, India; Pin-400 076; 918348690112 (M)

Sudipta Sengupta, E-mail: sudiptasg@yahoo.com Info: Department of Geological Sciences, Jadavpur University, Kolkata, West Bengal, India; Pin-700 032; 0910332457 2712 (O)

Hetu Sheth, E-mail: hcsheth@iitb.ac.in Info: Department of Earth Sciences, Indian Institute of Technology Bombay, Powai, Mumbai, Maharashtra, India; Pin-400 076; 9102225767264 (O)

Ichiko Shimizu, E-mail: ichiko@eps.s.u-tokyo.ac.jp Info: Department of Earth and Planetary Science, University of Tokyo, 7-3-1 Hongo, Bunkyo-ku, Tokyo 113-0033, Japan; 81-3-5841-4513

Toshihiko Shimamoto, E-mail: shima_kyoto@yahoo.co.jp Info: State Key Laboratory of Earthquake Dynamics, Institute of Geology, China Earthquake Administration, P. O. Box 9803, Beijing 100029, China; 18600262139 (M)

Kazuyuki Shiraishi, E-mail: kshiraishi@nipr.ac.jp Info: Geology Group, National Institute of Polar Research, 10-3 Midori-cho, Tachikawa, Tokyo 190-8518, Japan; Phone: +81-42-512-0603

Luiz Sérgio Amarante Simões, E-mail: lsimoes@rc.unesp.br Info: Universidade Estadual Paulista Júlio de Mesquita Filho, Institute of Geosciences and Exact Sciences of Rio Claro, Department of Petrology and Metallogeny, Av. 24-A, n° 1515, Bela Vista, CEP: 13506-900, Rio Claro, SP – Brazil; Tel: (55 19) 3526-9257

Guido Sibaja Rodas, E-mail: guisibro@gmail.com Info: San Isidro de Coronado. 50 m al este de la entrada al Restaurante Lone Star Grill, Condominio Quintana de los Reyes número 21. Vásquez de Coronado. San José, Costa Rica; (506) 2292-2296, (506) 8998-6018

Bikramaditya Singh, E-mail: rkaditya17@rediffmail.com Info: Wadia Institute of Himalayan Geology, GMS Road, Dehradun, Uttarakhand, India; Pin-248 001; 01352525103 (O)

Aabha Singh, E-mail: aabs_22@yahoo.co.in Info: Department of Geology, Fergusson College, F.C. Road, Shivaji Nagar, Pune, Maharashtra, India; Pin-411 004; 9920166093 (M)

Shailendra Singh, E-mail: singhgaur4@gmail.com Info: Geological Survey of India, GSITI, FTC Bhimtal, Northern Region, Lucknow, Uttarpradesh, India; 9450093216 (M)

Manuel Sintubin, E-mail: manuel.sintubin@ees.kuleuven.be Info: Geodynamics & Geofluids Research Group, Department of Earth & Environmental Sciences, KU Leuven, Celestijnenlaan 200E - box 2410, BE-3001 Leuven; Phone: +32 (0)16 32 64 47 - +32 (0)16 32 78 00

Ruth Soto, E-mail: r.soto@igme.es Info: Instituto Geológico y Minero de España (IGME), Unidad de Zaragoza. C/ Manuel Lasala, 44, 9B, 50006 Zaragoza, Spain; +34 976 555 153 (ext 26)

Frank Strozyk, E-mail: frank.strozyk@emr.rwth-aachen.de Info: EMR - Energy & Mineral Resources Group, Geological Institute, RWTH Aachen University, Wuellnerstr. 2, 52062 Aachen, Germany; Phone: +49-241-80-95718

Yutaka Takahashi, E-mail: takahashi-yutaka@aist.go.jp Info: Orogenic Process Research Group, Institute of Geology and Geoinformation, Geological Survey of Japan, AIST, 1-1-1 Higashi, Tsukuba, Ibaraki 305-8567, Japan; Phone: +81-29-861-3933

Tetsuhiro Togo, E-mail: duketogotetsu@gmail.com Info: Institute of Earthquake Volcano Geology, National Institute of Advanced Industrial Science and Technology (AIST), 1-1-1 Umezono, Tsukuba, Ibaraki 305-8568 Japan; Tel: 81-9099699526

Balázs Törő, E-mail: torobala@gmail.com Info: Department of Geological Sciences, University of Saskatchewan, Canada; +1 306 966 5737

Tsuyoshi Toyoshima, E-mail: ttoyo@geo.sc.niigata-u.ac.jp Info: Department of Geology, Faculty of Science, Niigata University, 8050 Ikarashi-2-nocho, Niigata 950-2181, Japan; +81-25-262-6199 (O)

Toshiaki Tsunogae, E-mail: tsunogae@geol.tsukuba.ac.jp Info: Graduate school of Life and Environmental Sciences (Earth Evolution Sciences), University of Tsukuba, Ibaraki, 305-8572, Japan; Phone: +81 29 853 5239

Janos L. Urai, E-mail: j.urai@ged.rwth-aachen.de Info: EMR - Energy & Mineral Resources Group, Geologie - Endogene Dynamik, RWTH Aachen University, Lochnerstrasse 4-20, D-52056 Aachen, Germany; Tel: +49-241-80-95723

Gianluca Vignaroli, E-mail: gianluca.vignaroli@uniroma3.it Info: Dipartimento di Scienze, Università degli Studi Roma Tre, L.go S.L. Murialdo, 1, 00146 Rome, Italy; +39 06 57338043 (M)

Simon Virgo, E-mail: s.virgo@ged.rwth-aachen.de Info: EMR - Energy & Mineral Resources Group, Geologie - Endogene Dynamik, RWTH Aachen University, Lochnerstrasse 4-20, D-52056 Aachen, Germany; Tel: +49-241-80-98438

Marko Vrabec, E-mail: marko.vrabec@geo.ntf.uni-lj.si Info: University of Ljubljana, Faculty of Natural Sciences and Technology, Department of Geology, Privoz 11, SI-1000 Ljubljana, Slovenia; 38612445412 (O)

Lu Yao, E-mail: yaolu_cug@163.com Info: State Key Laboratory of Earthquake Dynamics, Institute of Geology, China Earthquake Administration, P. O. Box 9803, Beijing 100029, China; Tel: 86-13426054959

Ran Zhang, E-mail: Ran.Zhang@bhpbilliton.com Info: BHP Billiton Petroleum, 1360 Post Oak Blvd., Ste. 150, Houston, TX 77056-3030 USA; +1 713 297 6568

Preface

Documentation of structures in different scales is the first step in many structural geological studies. This edited atlas gives an overview of diverse structures. Due to lack of space or inappropriateness, sometimes interesting structural snaps cannot be published in journals. This book fills that gap.

Acknowledgments

Thanks to Mohanapriyan Rajendran, Priya Srikumar, Marisa LaFleur, Amy Shapiro, Louisa Hutchins, and John Fedor (Elsevier) for editing, and to all the contributors and reviewers. Philippe Herve Leloup, Chris Talbot and an anonymous reviewer are thanked for reviewing the book proposal and for providing positive comments. Research students and teaching assistants helped and I thank them: Tuhin Biswas, Narayan Bose, Achyuta Ayan Misra, Aninda Ghosh, Rajkumar Ghosh, Dripta Dutta, Uddipan Das, and many others. Thanks to my wife Payel Mukherjee for her patience.

Chapter 1

Folds

KEYWORDS
Folds; Folds not related to shear zones; Overturned fold; Shear zone related fold; Sheath fold; Superposed fold.

Two of the most intensely studied aspects in structural geology are morphology and genesis of folds (see Ramsay, 1967; Hudleston and Lan, 1993; Ez, 2000; Harris et al., 2002; Harris, 2003; Alsop and Holdsworth, 2004; Mandal et al., 2004; Carreras et al., 2005; Bell, 2010; Hudleston and Treagus, 2010; Godin et al., 2011; Harris et al., 2012a,b; Llorens et al., 2013; Mukherjee et al., in press). Of particular importance is whether folds found inside ductile shear zones are related to ductile shear (e.g., Mandal et al., 2004; Carreras et al., 2005; Bell, 2010). This chapter presents folds of different geometries and generations, some related with ductile shear zones, from different scales (Figures 1.1–1.87).

FIGURE 1.1 **Upright folds and folded boudins resulting from continental collision of East and West Gondwana.** The boudins of dark-colored amphibolite (Fb) in light-colored biotite-hornblende gneiss have originally pancake shapes with flattening parallel to compositional layering of gneiss, and resulted from the layer-parallel extension and thinning of crustal rocks within 640–600 Ma (Toyoshima et al., 1995). The folds with wavelengths of 20–30 m are parasitic upright folds of larger-scale upright fold related to 600–560 Ma sinistral transpression and crustal shortening during the collision (Toyoshima et al., 2013). The boudins (Fb) folded by the parasitic folds suggest that the tectonic regime changed from layer-normal to layer-parallel compression (Toyoshima et al., 2013). Osanai et al. (2013) presented SHRIMP U–PB ages for metamorphic rocks from the Sør Rondane Mountains, East Antarctica, and recognized periods of ultrahigh-temperature metamorphism (pre-main metamorphic stage) during 750–700 Ma and granulite- to amphibolite-facies metamorphism during 640–600 Ma. Location: 72°09′42″S, 25°31′50″E, the southern part of Salen in the Sør Rondane Mountains, East Antarctica. *(Tsuyoshi Toyoshima, Masaaki Owada, Kazuyuki Shiraishi)*

Atlas of Structural Geology. http://dx.doi.org/10.1016/B978-0-12-420152-1.00001-6

FIGURE 1.2 **Fault-bend-fold viewed along NNW-SSE section within Tethyan Himalayan succession.** Compression in the Tethyan Himalaya indicated. Width of view: ~80 ft. The fold is developed within thinly bedded, presumably less viscous, argillaceous rocks of Lower Cretaceous age. North vergent folding in the Tethyan Himalaya resulted from strong drag induced by channel flow extrusion of the Greater Himalaya at south, as Godin et al. (2011) proposed from Nepal Himalaya. Note opposite senses of normal drag (Grasemann et al., 2005; Mukherjee, 2007; Mukherjee and Koyi, 2009; Mukherjee, 2010, 2011a, 2014a,b) across the fault plane. Latitude 32°19′54″N, Longitude 78°0′29″E. Kibber village, Spiti, Himachal Pradesh, India. *(Mainak Choudhuri)*

FIGURE 1.3 **Tri-shear fault-propagation folding.** Greater Himalayan gneissic rocks. The fold axis trends N330°. Location: 28°32′48.61″N; 94°13′47.24″E. Scale: Hammer (white circled). *(George Mathew, Dnyanada Salvi)*

FIGURE 1.4 **(a) Full section and (b) zoomed into an excellent outcrop-scale example of an "out-of-syncline" fold accommodation fault** (Mitra, 2002; Deng et al., 2013), Triassic limestones, Spiti valley, near Losar village, Himachal Pradesh, India. Fold accommodation faults accommodate strain on the fold limbs. Here the synform is present to the SW. Its NE limb accommodated strain. Location: 32°23′47.63″N, 77°57′2.09″E. *(Achyuta Ayan Misra)*

FIGURE 1.5 **Fault-bend fold.** The fault plane is not perfectly straight. Presuming a "normal drag," reverse faulting deciphered. Thinly bedded meta-siltstone. 15-mm pencil for scale. Location: Chail Group, SE of Shimla, Himachal Pradesh, India. *(Subhadip Mandal)*

FIGURE 1.6 **The Eocene limestone–shale sequences of the Siang fold-and-thrust belt, Arunachal Pradesh, India.** The sequence is bounded by parallel and arcuate thrusts at N (Acharyya and Saha, 2008). The southward propagating Eocene sequence is thrusted and folded. Here the fold has its right limb steeper than the left one. The right limb rotated dextrally and is traversed by two basic dykes (BV). The left dyke is also slipped dextrally, the middle part got boudinaged and the lower part dislocated creating a subvertical fault. Since the two dykes follow surfaces of opposite movement, the sequence within the dykes becomes horses between the floor and the roof thrust (McClay and Insley, 1986) with top-to-right shear. The horses are sigmoidal. Thick white broken lines: axial trace of the fold and one sigmoid. L.St.: Limestone, Sh: shale, BV: Basic volcanics. Lower Siang district, Arunachal Pradesh, India. N28°10′8″, E95°12′7″. Several Neogene N-trending rifts present in the Eastern Himalaya (Yin, 2006). The geometry of the Siang fold thrust belt is not studied so far considering the orientation and extension of the rifts in the E Himalayan Syntaxis. *(Tapos Kumar Goswami)*

FIGURE 1.7 **Spectacular exposure of a thrust fault and related footwall syncline at the summit of Mt Prena belonging to the E-W-trending Gran Sasso thrust system in the Central Apennines fold-and-thrust belt of Italy.** The thrust emplaces the massive paleo-platform dolostones of the Upper Triassic Dolomia Principale Formation, in the hanging wall, onto the well-bedded slope-to-basin calcarenites and pelagic limestones of the Middle-Jurassic–Lower Cretaceous Calcari Bioclastici inferiori and Maiolica Formations, in the footwall, which are involved in a close overturned syncline. Further details about the structures of the Gran Sasso thrust system can be found in Calamita et al. (2004, 2009) and Satolli et al. (2005). Location: view of Mt Prena from the W (42.449190° 13.651570°) ~6 km S to the Isola del Gran Sasso d'Italia village, province of Teramo (Italy). The Mt Prena (2561 m.a.s.l.) is one of the highest peaks among the summits in the Gran Sasso d'Italia, which is the highest mountain of the Apennines. Width of view ~2 km. *(Paolo Pace)*

FIGURE 1.8 **Folded strata in the Tibetan Himalaya.** Rock consists of shale, meta-sandstone and carbonates. Photo view toward NW. Location: Near Kagbeni, Tibetan Himalaya, Kali-Gandaki section, Central Nepal Himalaya. *(Subodha Khanal)*

FIGURE 1.9 **Recumbent fold in mesoscale.** Photo at Ordesa and Monte Perdido National Park (Spanish Pyrenees) to the Northeast, in Llanos de Salarons. 30T 740329.94 E 4728732.43 N. Paleocene and Eocene limestones (Salarons- and Gallinera Formations) in gray and overlying Eocene turbidites (Hecho Group) in brown (Séguret, 1972; Ternet et al., 2008). *(Ruth Soto)*

FIGURE 1.10 **Drag fold (recently reviewed in Mukherjee 2014) produced by faulting within Gulcheru Quartzite.** Quartzite is of Paleoproterozoic age. SSW of Vempalli village, Andhra Pradesh, India. *(Atanu Mukherjee)*

FIGURE 1.11 **The Ayishan detachment lies between the Karakoram fault on its E and the Great Counter thrust on its W, in West Tibet.** The detachment mantles two NW-SE elongated large doubly plunging gneiss domes. The detachment consists of low-angle top-to-SE extensional faults that overlie a thick top-to-SE shear zone, which juxtaposes Cretaceous-early Tertiary granitoids in its hanging wall against mylonitic rocks in its footwall. Both the faults and underlying shear zone folded about an axis that parallel the strike of the range. See Wang et al. (2014) for detail of tectonics. *(Ran Zhang)*

FIGURE 1.12 **Kink folded intercalated layers of shales and limestones bound by a SE dipping detachment/décollement at bottom.** Note (1) different geometries of folds along the fold train, (2) folding varies in style from intrados up to the extrados, and (3) fracture along axial traces. Lesser Himalaya, N31°14.578′, E76°58.96′, Mangu village, district: Solan, Himachal Pradesh, India. *(Tuhin Biswas)*

FIGURE 1.13 **Folded "khaki" green shale at Jitpur (India), Talchir Formation.** Initially folded layers later brittle deformed producing two joints sets at high angle. Thus needle-shaped shale formed. *(Ananya Basu)*

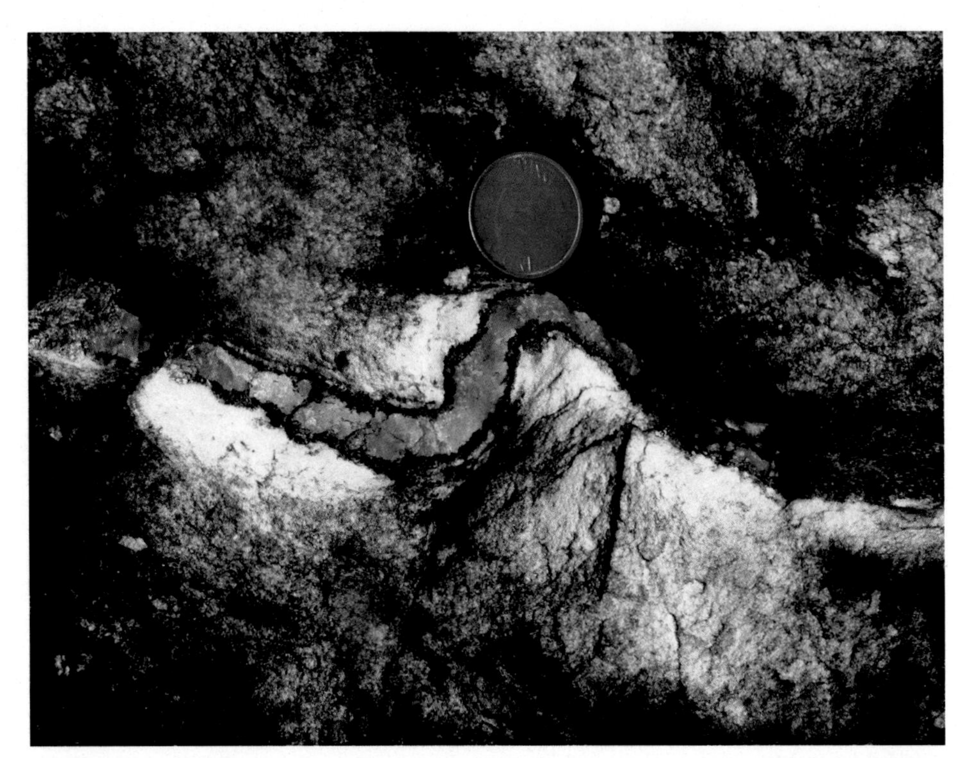

FIGURE 1.14 **A folded quartz vein: tremolite grew along its boundary.** Surrounding rock: dolomite. N23°07.742′ E79°48.809′ to N23°07.563′ E79°48.665′. Bhedaghat, Jabalpur, Madhya Pradesh, India. *(Sreejita Chatterjee)*

FIGURE 1.15 **Cuspate folds of quartz vein within meta-greywacke.** Kumbhalgarh Formation, ~85 km NW of Udaipur, Rajasthan, India. *(Swagato Dasgupta)*

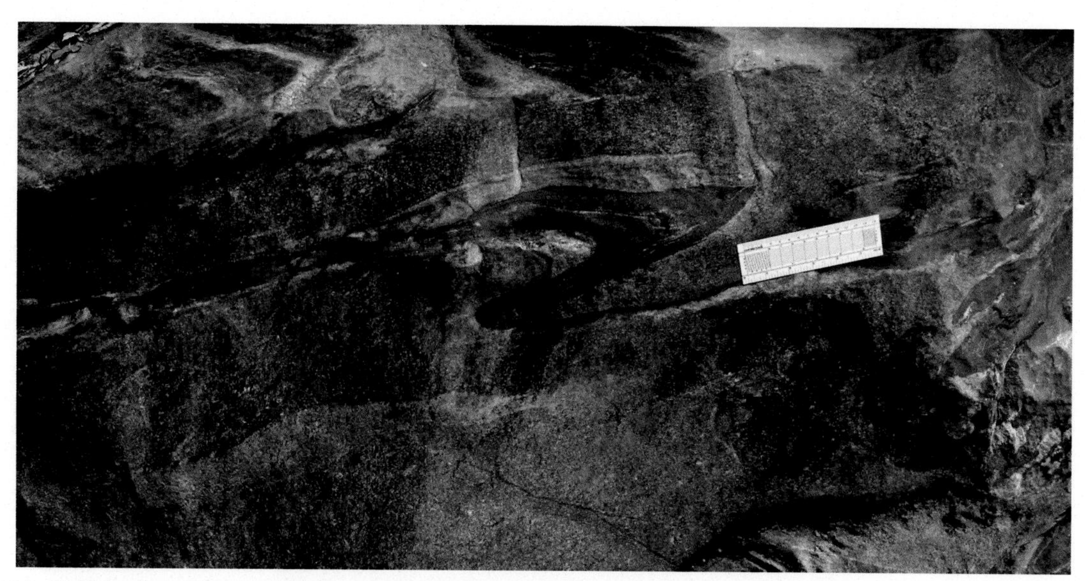

FIGURE 1.16 **Folded migmatitic rock.** Melanosome layer is overturned folded with arrow-head hinge. Competency contrast between the melanosome layer and the remaining portion of the rock probably is one of the key controlling parameters of such folding. Maithon, West Bengal, India. *(Ananya Basu)*

FIGURE 1.17 **Intensely deformed lower limb of tight isoclinal folds in highly foliated garnetiferous gabbroic rocks.** The lower limb shows a variety of folds and a series of small-scale brittle-ductile shear zones. All the folds show gentle to moderate plunge toward S/SE. The shear zones strike ~E-W and often associate with flanking structures of negative slip. Those are slipped sinistrally. The magnitude of slip increases forming arcuate shear zone and that finally coincide with the axial plane of the fold. On the right side of the photograph, notice the presence of folded quartzofeldspathic layers with sinistral displacement along the axial plane and stretching of one of the limbs. The outcrop is located in the NW part of E-W trending closed structural form of Kanjamalai Hills (~5 km W of Salem). These structures form a part of detachment zone, which is characterized by complex folding with predominant tight/isoclinal folds with variable trend, limb rotations, and the hinge line variations often leading to lift-off like fold geometries and deformed sheath folds. See Drury and Holt (1980), Gopalakrishnan (1994), Chetty (1996) and Santosh et al. (2009) for regional geology and tectonics. *(T.R.K. Chetty)*

FIGURE 1.18 **A chevron fold with planar limbs and sharp/pointed hinge.** Planar subvertical axial plane. *From quartzite, Ajmer, Rajasthan, India. (Moloy Sarkar)*

FIGURE 1.19 **Large-scale folded banded magnetite quartzite and magnetite garnet biotite schist intercalation.** Tiranga hill, Pur area, Bhilwara, Rajasthan, India. *(Ranjan Gupta)*

FIGURE 1.20 **Folded banded iron formation of alternate layers of ferrugineous and quartzose materials.** Chitradurga district, Karnataka, India. *(Aabha Singh)*

FIGURE 1.21 **Photograph from the Chrystalls Beach Complex, coastal Otago, New Zealand, showing folded interlayered sandstones (white) and mudstones (dark gray).** The rocks form part of the Otago schist, related to Triassic–Jurassic subduction and accretion at the Gondwana-Paleopacific margin (e.g., Mortimer, 2004). The Chrystalls Beach Complex is an accretionary mélange, where stratal disruption occurred progressively from early diagenesis through to chlorite–actinolite subgreenschist facies conditions (Nelson, 1982). The fold in the photograph illustrates that more rigid sandstone layers are both disrupted by layer-parallel extension, and thickened by folding and small-scale thrusts. Mudstone appears to have filled gaps in the sandstone layer by ductile flow, but tensile fractures in the sandstone attest to local brittle deformation. This mixture of continuous and discontinuous deformation is typical of this tectonic mélange. The bulk rheology of the mélange may therefore depend on the relative abundance of low-viscosity mudstone and more rigid sandstones; ductile deformation at the mesoscale, as seen in the photograph, likely occurs when mudstone dictates the bulk rheology, but brittle deformation may prevail if sandstone forms a rigid framework (Fagereng and Sibson, 2010). *(Ake Fagereng)*

FIGURE 1.22 **Rhythmic occurrence of the relatively undeformed massive sandstone turbidite beds within the background of calcareous mudstone units (pelagic-hemipelagic sediments and "low-density" turbidites), which contain folded and closely spaced vertical joint surfaces.** The flexural slip folding of joints with bedding-parallel axial surfaces is caused by compaction of the mudstones within sedimentary diagenetic regime. The sandstone beds contain intermittent and rare bedding-perpendicular joints, which are not folded. This implies distinctly different postdepositional rheologic histories for mudstones and sandstone. After deposition, the sediments underwent unidirectional extension along the slope – preferably more in mudstone than in sandstone beds, thereby forming the bedding-perpendicular joint surfaces. Unlike in sandstones, the joint surfaces in mudstones subsequently were folded by further compaction. (Figure 1.23) Closer view of the deformed mudstone. The facies-tract assemblage is possibly of an unconfined deep marine slope apron. From late Eocene–early Oligocene Ceylan Formation, Thrace Basin, Gelibolu Peninsula, North-West Turkey. *(Sudipta Dasgupta, Luis A. Buatois)*

FIGURE 1.23

FIGURE 1.24 **Angular folds (kink bands) in tourmaline-bearing garnet-mica schist.** Dark green and pale brown minerals in dark bands are of tourmaline. Small dark green crystals are the curved three-sided (basal) cross sections of the tourmaline. The folds are asymmetric showing larger limbs in the left side. In the hinge zones some prismatic sections of tourmaline are slightly bent but most of them are undeformed. Bent tourmaline crystals can be interpreted as pre- to syn-folding, whereas undeformed tourmaline grew at least partially after folding. Observed abundant basal sections suggest that tourmaline also grew parallel to the fold axis orientation. Section normal to the fold axis and to the foliation. Width of view 22 mm. Plain polarized light. Location: Sierra de Ambato, Catamarca Province, NW Argentina. GPS point (WGS84): S 27°38′35.6″ - W 66°12′39.5″; Rock type: tourmaline-bearing garnet-mica schist; Formation name: Quebrada del Molle Metamorphic Complex; Age (relative): Ordovician; Facies/grade: greenchist facies / garnet zone. *(Mariano A. Larrovere)*

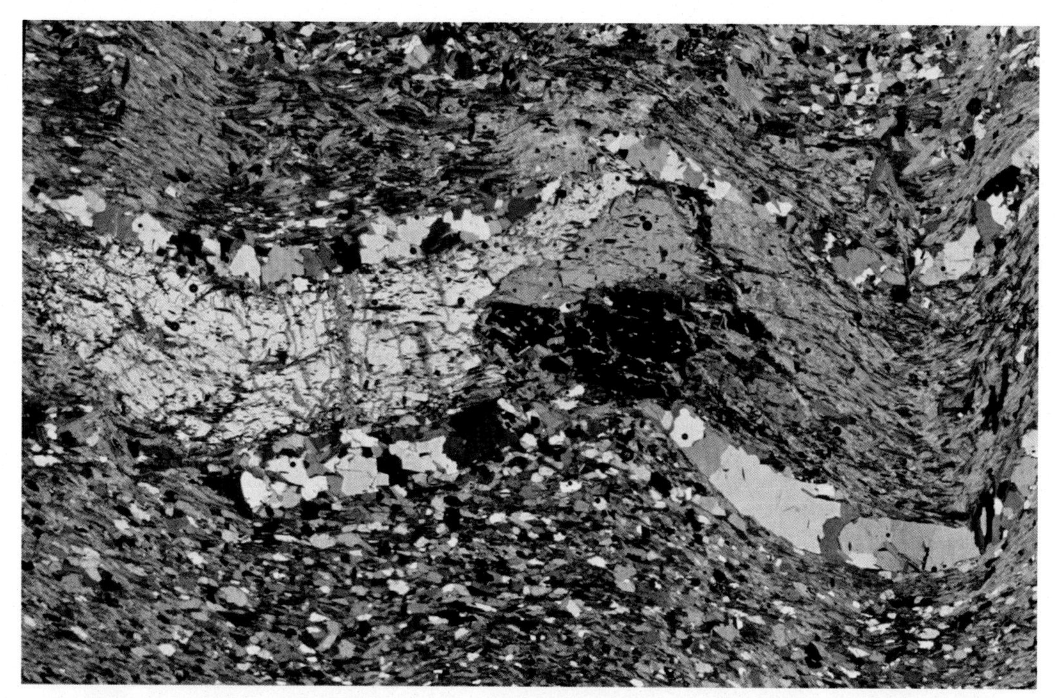

FIGURE 1.25 **Microfolded porphyroblast of cordierite (in the center of the photo) in gneiss.** Subgrains occur within the large elongate deformed cordierite grain. They are recognized optically by low relief boundaries and differential extinction between subgrains. Inside the cordierite crystal is observed a bent inclusion pattern marked by micas (internal foliation) that shows continuity with the folded subhorizontal gneissic foliation S_1. The crystal of cordierite is bordered by two ribbons of recrystallized quartz that also are folded. A subvertical crenulation cleavage S_2 can be recognized. The cordierite porphyroblast may be classified as pretectonic with respect to S_2 since cordierite is folded and internally deformed by the same deformation phase (D_2) that generated the crenulation cleavage S_2. On the other hand, cordierite porphyroblast is interpreted as post-tectonic with respect to the subhorizontal gneissic foliation since it includes S_1. Section normal to the microfold axis and to the foliation. Width of view 13 mm. Cross polarized light. Location: Sierra de Aconquija, Catamarca Province, NW Argentina. GPS point (WGS84): S 27°34′07.6″ - W 66°14′20.6″; Rock type: cordierite-mica gneiss; Formation name: Quebrada del Molle Metamorphic Complex; Age (relative): Ordovician; Facies/grade: greenchist facies / medium grade. *(Mariano A. Larrovere)*

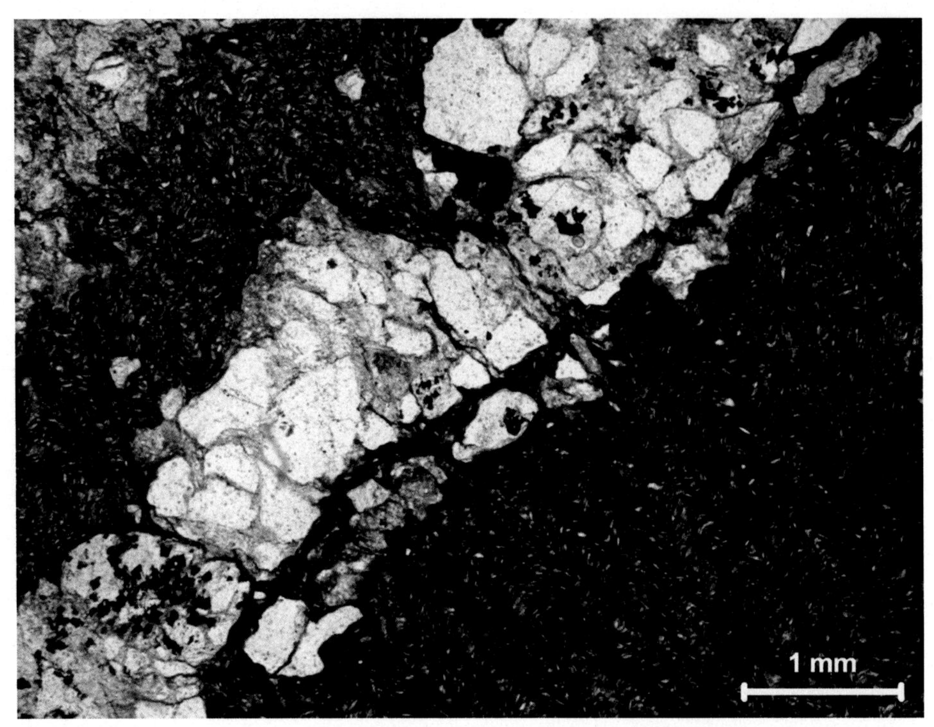

FIGURE 1.26 **Crenulation cleavage.** Crenulation cleavage (Hobbs et al., 1976, p. 217) in thin pelitic laminae intercalated with sandstone beds. A bedding parallel, phyllosilicate-rich fabric (due to load metamorphism) in the pelitic laminae is crenulated, while the sandy laminae remain straight, but cut by occasional pressure solution seams grading into crenulation cleavage seams in the pelitic laminae. Quartz grains in the sandy laminae retain there original angular to subangular shape. PPL photomicrograph, Bowmore Sandstone, Rhinns of Islay, Scotland, bar scale 1 mm. PPL, plain polarized light. *(Dilip Saha)*

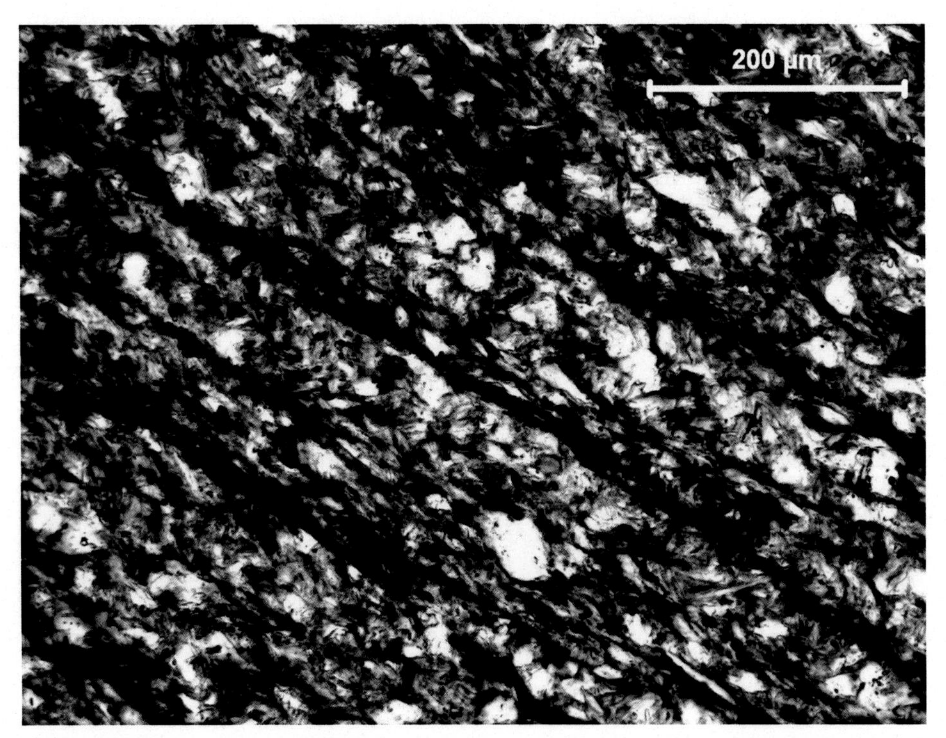

FIGURE 1.27 **Disjunctive spaced cleavage in meta-siltstone.** Note (1) narrow cleavage seams, (2) thin dark bands (P-domain: Williams et al., 2001) dipping to the right, and (3) alternating with microlithons (Q-domain). Q-domains are characterized by randomly oriented chlorite and muscovite flakes. P-domains have tiny subparallel phyllosilicate flakes and concentration of ferruginous opaques. The distributed quartz grains also show poor shape preferred orientation. Photo in plane polarized light. Colonsay Group, western Islay, Scotland. Bar scale: 200 μm. *(Dilip Saha)*

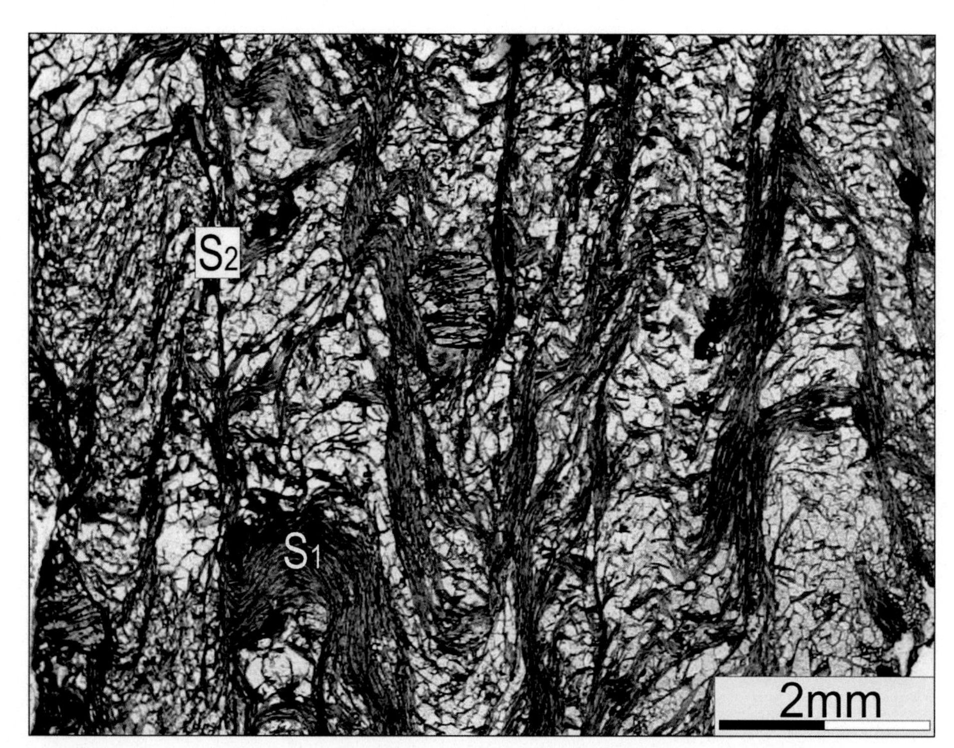

FIGURE 1.28 **Photomicrograph under the cross-nichol of Mesoproterozoic garnetiferous mica schist, upper unit of Lesser Himalayan Crystallines, Kameng valley, Western Arunachal Himalaya showing latter schistosity/crenulation cleavage (S_2) from early crenulation (S_1) that occurred as microfolds in quartz-rich microlithons.** GPS location: N27°17′11.3″; E92°15′48.3″ (Bell and Rudenach, 1983; Passchier and Trouw, 1996; Bikramaditya Singh and Gururajan 2011). *(Bikramaditya Singh)*

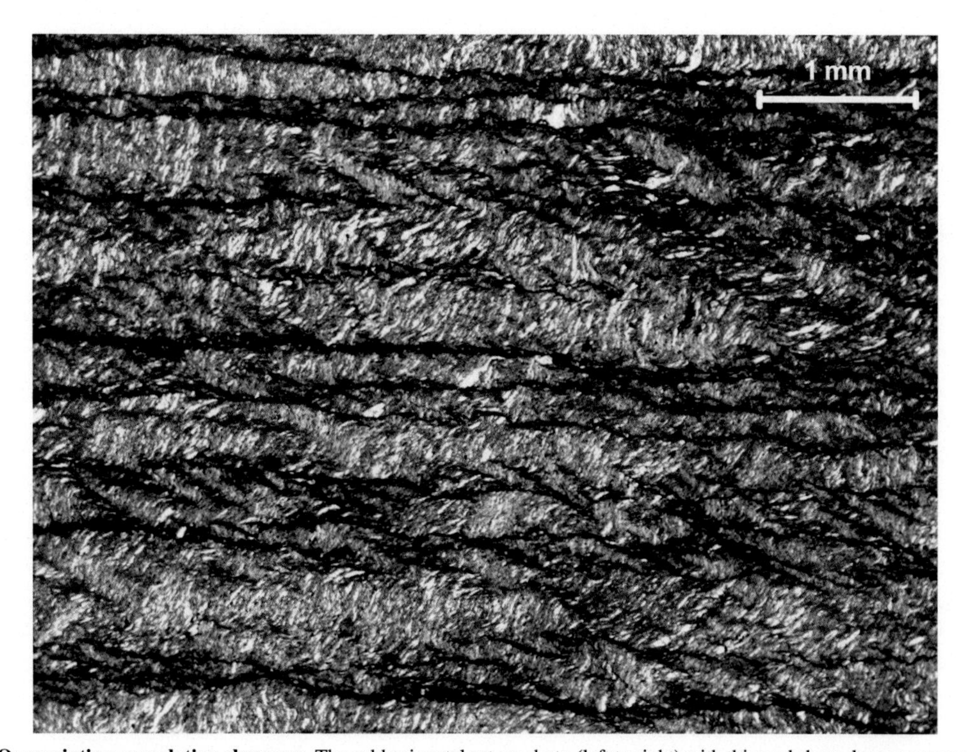

FIGURE 1.29 **Overprinting crenulation cleavages.** The subhorizontal set on photo (left to right) with thin and sharp, longer seams transpose an early phyllitic cleavage (slaty cleavage) defined by alignment of chlorite and muscovite flakes, and flattened quartz grains. A still later set of crenulations affect both the phyllitic cleavage and the subhorizontal crenulation cleavage seams, leading to a second crenulation cleavage with shorter narrow seams gently dipping to the right of photo. PPL photomicrograph, phyllites of Colonsay Group, Saligo Bay, Western Isaly, Scotland, bar scale 1 mm. PPL, plain polarized light. *(Dilip Saha)*

FIGURE 1.30 **Similar folds within the Los Cabos Series quartzites (Ordovician), below the basal thrust of the Mondoñedo nappe.** Folds show characteristic thickened hinges and the left limb shows top-to-right (up) shear. Left hand fold is approximately 2 m wide at base. West Asturian Leonese Zone, Tapia de Casariego, Spain. *(Amy Ellis)*

FIGURE 1.31 **Disturbed layer in lacustrine organic rich dolomitic lime mudstone/"oil-shale."** Eocene Green River Formation, Piceance Creek Basin, Colorado, USA. The deformed interval shows a lower, brittle-ductile part, with fault-propagation fold, and a mixed, fragmented interval above. The pronounced difference in the style of deformation reflects different rheological properties of the two sedimentary packages during deformation. Sediments of the lower part were in a semilithified state, while sediments above were softer. Undulation of the upper boundary might indicate deposition after the deformation event, which took place at the sediment–water interface. Deformation structures with similar characteristics have been previously described as "mixed layers" in the literature (e.g., Marco and Agnon, 1995; Rodríguez-Pascua et al., 2000), showing gradual upward transition in the style of deformation induced by seismic shaking. The depositional environment of the host sediments, with very gentle or flat lake floor also point to a tectonic origin. *(Balázs Törő, Brian R. Pratt, Sudipta Dasgupta)*

FIGURE 1.32 **Parasitic minor folds exhibiting an S-shaped sense of asymmetry within the overturned eastern forelimb of the NNW-SSE-trending Montagna dei Fiori anticline in the Central-Northern Apennines foreland fold-and-thrust belt, Italy.** Folding involves the well-bedded pelagic limestones of the Upper Cretaceous–Eocene Scaglia Rossa Formation. The contrasting lithological response to folding is exposed: thick and stiff calcarenitic layer remains undeformed, whereas the thinly layered mudstone strata are remarkably parasitic S-shaped folded. Calamita et al. (1998) and Scisciani et al. (2002) described the structures of the Montagna dei Fiori anticline. Calamita et al. (2011, 2012a) detailed thrust-related folding mechanisms and examples from the Central-Northern Apennines. Location: Sant'Angelo Caves (42.752752° 13.623099°) W to the Ripe di Civitella village, province of Teramo (Italy). Width of view ~16 m. *(Paolo Pace)*

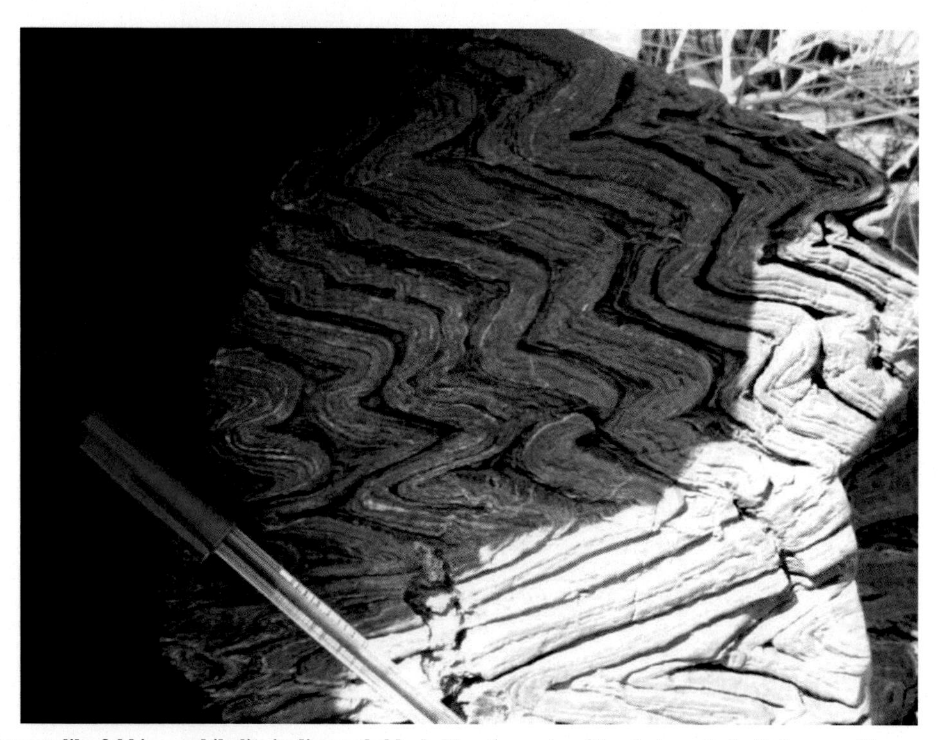

FIGURE 1.33 **Chevron-like fold in amphibolite in dismantle block.** Near the center of the photograph, the limb snapped in one case. Akjoujt, Inchiri region, Western Mauritania. *(Rajib Sadhu)*

FIGURE 1.34 **A tight isoclinal fold, with uneven thickness of limbs and hinge regions, in amphibolite.** Akjoujt, Inchiri region, Western Mauritania. *(Rajib Sadhu)*

FIGURE 1.35 **Intrusion of nearly straight quartz veins in the folded quartzite intercalated with mica schist.** Notice fold hinges in few cases are thicker than the limbs. Ghatshila, Singhbhum, Jharkhand, India. *(Ranjan Gupta)*

FIGURE 1.36 **Cross-polarized photomicrograph of a folded scapolite porphyroclast partially wrapped by quartz ribbons and in a matrix of grain-size reduced scapolite and alkali feldspar.** This grain occurs in a syn-deformational, late-Grenville-aged vein from the NW Adirondack Mountains (New York, USA). It demonstrates that although porphyroclasts are typically considered to be more competent than their surrounding matrix phases, they may accumulate significant internal strain. This is similar to folded quartz grains reported in Mukherjee (2010), Mukherjee and Koyi (2010) and Mukherjee (2013). The pronounced internal folding and dynamically recrystallized grain margin indicates that scapolite was only marginally more competent than alkali feldspar at the conditions of deformation experienced by this vein. *(Chloë Bonamici)*

FIGURE 1.37 **Asymmetric folds with stretched quartz boudins in the metasedimentary rock of Higher Himalayan Crystallines, Tawang valley, Western Arunachal Himalaya.** GPS location: N27°39′30.5″; E91°52′17.2″. *(Bikramaditya Singh)*

FIGURE 1.38 **Tightly folded quartz vein in dolomitic host rock, Central Indian Suture Zone, Jabalpur, Madhya Pradesh, India.** *(Jyotirmoy Paul)*

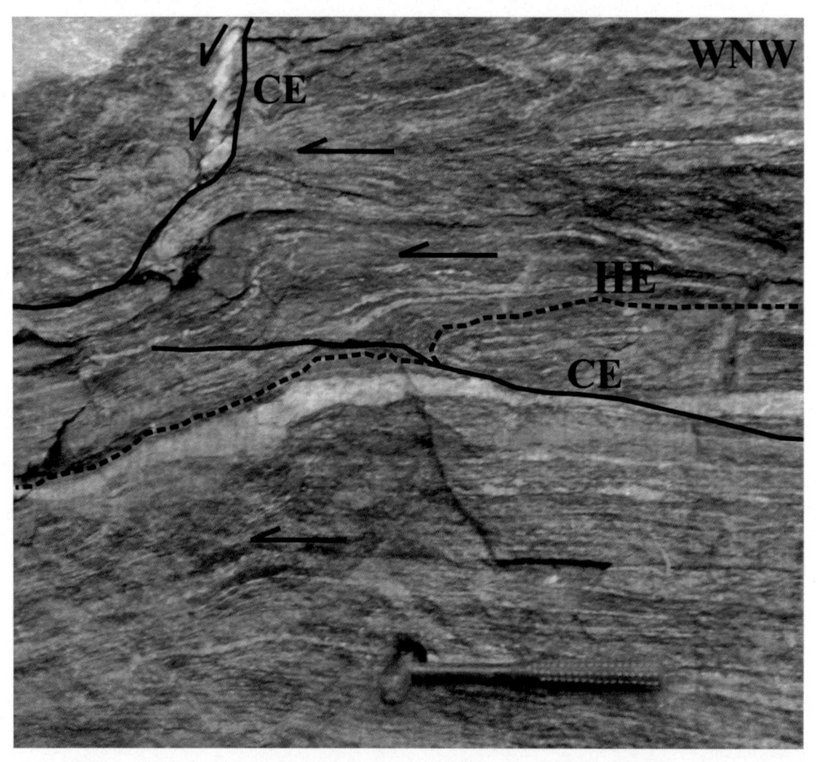

FIGURE 1.39 **The Higher Himalaya gneisses near Tato, Siyom River section, West Siang district, Arunachal Pradesh.** These garnetiferous gneisses have the general trend of WNW-ESE dipping at low angle toward N. These gneisses are migmatitic near Tato and represent a high strain ductile shear zone. Flanking folds (Passchier, 2001; Mukherjee, 2014b) in these high-grade gneisses represent sinistral thrusting. The host element is the compositional layering and the cross-cutting elements (CE) are a fault in the central part while in the top left a subvertical vein which are responsible for flanking of the host layers. The entire fold train is deflected along the axis. The thick quartz vein in the central part post dates the flanking of the host layers and thus shows zero slip. The flanking folds in the central part clearly indicate positive slip, lift, and over roll. The lower mylonitic part shows limbs of the F_1 folds stretched and boudinaged. The photograph is unique in the sense that two CE,i.e., a fault in the central part and a subvertical vein in the top left part both are responsible for the flanking of the host fabric elements. The shear sense in the top vein is at high angle to the senses in the central part. The thick quartz vein in the central part in a way dividing the area into one part showing flanking structures while the lower part with mylonitic fabric. West Siang district, Arunachal Pradesh, India. Coordinate: N28°30′976″, E94°22′642″. *(Tapos Kumar Goswami)*

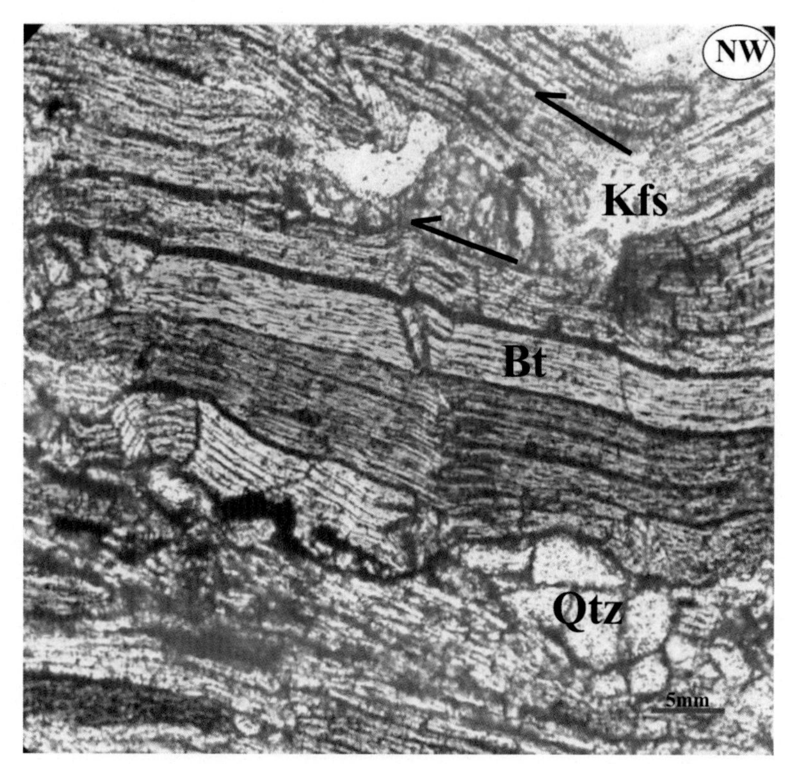

FIGURE 1.40 **Micaceous Dafla sandstone of Middle Siwalik (N27°05′057′, E92°35′ 258′), of Kameng River section, West Kameng district, Arunachal Pradesh, India.** The bottommost Dafla sandstone thrust over the Subansiri Formation near Tipi. Kink bands are frequent in the Dafla sandstones. These bands in sandstones formed at relatively low temperature and must be related to local stress. The kink bands vary from simple to very complex type and verges in the same direction of the outcrop-scale folds. Phyllosilicates deform homogeneously when shear stress applied to the basal planes (001). However, in the nonbasal planes the applied shear stress produces kinks or fractures (Kronenberg et al., 1990; also see Van Loon et al., 1984). Syndiagenetic deformation achieve ~150 °C (Lin, 1997). Brown and green (chloritised) mica layers in Dafla sandstone shows kink bands of different styles and verge toward SE. Both mobile and fixed hinge kinking are observed in the central part. The top mica layer shows the long limb above the shorter limb and is a thrust. The left central part shows a curved kink boundary indicating differential movement of the basal plane front. *(Tapos Kumar Goswami)*

FIGURE 1.41 **Z-type of buckle fold of amphibole-rich layer in metamorphosed granitic host rock. Chotonagpur Gneissic Complex, Maithon, West Bengal, India.** *(Jyotirmoy Paul)*

FIGURE 1.42 **Small-scale folds in banded gneiss in the hanging wall of the Main Central thrust, Photo view toward NW.** Axial trace of folds are parallel to the pen. Notice both rounded and sharp hinges of folds. Location: Unit I, Greater Himalaya, Kodari-Tatopani section, central Nepal Himalaya. For location and geology of the Main Central Thrust in the area, see Khanal et al. (2014). For review of tectonics of Main Central thrust, see Mukherjee (2013a,b). *(Subodha Khanal)*

FIGURE 1.43 **Crenulation cleavage in garnetiferous banded gneiss in the hanging wall of the Main Central thrust.** Penny for scale. Photo view toward NW. Location: Unit I, Greater Himalaya, Budhi-Gandaki section, Central Nepal Himalaya. *(Subodha Khanal)*

FIGURE 1.44 **Intrafolial folds with top to the left sense of shear in the banded gneiss.** Leucogranites are parallel to foliation and thins out toward top left. Location: Unit I, Greater Himalaya, Marsyangdi section, Central Nepal Himalaya. *(Subodha Khanal)*

FIGURE 1.45 **A gently inclined folded quartzite. Parasitic fold more prominent in the lower limb.** Ajmer district, Rajasthan, India. *(Moloy Sarkar)*

FIGURE 1.46 **Recumbent fold plunging towards E in hornblende gneiss.** Upper Aravalli Group. East of Bagdunda, Udaipur district, Rajasthan, India. *(Shruthi Narayanan)*

FIGURE 1.47 **Plunging fold in ferrogenous sandstone with alternate iron-enriched layer and silica layer.** Axial plane dips toward left. Akjoujt, Inchiri region, Western Mauritania. *(Rajib Sadhu)*

FIGURE 1.48 **Sheath fold.** Chitradurga district, Karnataka, India. *(Aabha Singh)*

FIGURE 1.49 **A folded layer of chert with closures at two directions.** Is this a sheath fold? Note layer ~ perpendicular fractures confined inside the folded chert body. *(Anupam Samanta)*

FIGURE 1.50 **Isoclinally folded chert layers with closures in two directions.** In few cases hinges are thicker than the limbs. *(Anupam Samanta)*

FIGURE 1.51 **Field photograph shows a boudinaged sheath fold.** The horizontal view is perpendicular to the mylonitic foliation with a downdip stretching lineation. The rock is a calcareous unit intercalated with quartzofeldspathic layers. The quartzofeldspathic layers show excellent development of sheath folds. Boudinage structure can also be seen in the upper quartzofeldspathic layers. Presence of boudins in these layers in a calcareous matrix indicates a strong competence contrast between them. Development of boudins in a section perpendicular to the stretching direction indicates that the deformation was of the flattening type and there is a component of compression across the shear zone. The rock is from the Phulad Shear Zone (Ghosh et al., 1999, Sengupta and Ghosh, 2004) which demarcates the western margin of Delhi Fold Belt, Rajasthan, India. The white bold arrow indicates geographic north. The width of the photo is 23 cm. Sample location is 25°39′46.1″N, 73°48′36.3″E. *(Sadhana M. Chatterjee, Sudipta Sengupta)*

FIGURE 1.52 **Doubly plunging round hinge isoclinals fold in calc-silicate.** The thinned limb in the photo center possibly indicates a top-to-left shear. Note orthogonal thickness varies along the limb. At one place, but not in every cases, the hinge region is thicker than the limb. Central Indian Suture Zone. Jabalpur, Madhya Pradesh, India. *(Jyotirmoy Paul)*

FIGURE 1.53 **Planar view of complex folding in intensely foliated high-grade quartzofeldspathic gneisses.** A well-defined curvilinear hinge zone is distinct in the central part followed by complex fold styles on either side suggesting that the rocks were sheared intensely. The variation in the hinge line indicates sheath geometry. All the folds exhibit varying plunges from near horizontal to steep values. The axial planes trend N-S and dip steeply toward E. Deformed sheath fold structures involving gabbroic rocks are also recorded in the vicinity. These structures form a part of detachment zone located in the W part of Kanjamalai Hills (Mukhopadhyay and Bose, 1994). High-pressure granulites have been described from the region (Saito et al., 2011; Anderson et al., 2012). Kanjamalai Hills occur ~5 km W of Salem along the Moyar–Bhavani Shear zone, the northern boundary of Cauvery Suture Zone, Southern Granulite Terrane (Ghosh et al., 2004). *(T.R.K. Chetty)*

FIGURE 1.54 **Neutral folding where the quartzose layers vary in thickness.** Hinge area in few cases show M-geometry. Chitradurga district, Karnataka, India. *(Aabha Singh)*

FIGURE 1.55 **Rotation of F_2 fold hinges near the Bastar Craton–Rengali Province boundary shear zone.** F_2 fold hinges preserved locally in quartzofeldspathic gneisses of the Bastar Craton, and rotated into parallelism with a late S_5 shear foliation related to the contact with the Rengali Province. At a distance from the shear zone, F_2 fold axes plunge steeply, but rotate into shallow plunges near the shear zone (the Kerajang Fault of Crowe et al., 2003). These F_2 folds are now noncylindrical, and have curved intersection lines with the S_5 shear plane. On National Highway-6, N of Padiabahal, Sambalpur district, Odisha, India. *(Arindam Dutta, Saibal Gupta, M.K. Panigrahi)*

FIGURE 1.56 **This ~10-m wide view shows the lower half of a major recumbent fold in a W looking sectional view across the Palghat Cauvery Shear Zone, South India.** The rock types include well-foliated mafic–ultramafic rock assemblages along with thin layers of quartzofeldspathic material (Yellappa et al., 2010). The rocks show near horizontal plunges and gentle to moderate dips to north. Pegmatite veins intruded these rocks both along and across the foliation planes. The outcrop is located near Anniyapuram along a N-S section between Namakkal and Mohanur that coincides the S boundary zone of the Cauvery Suture Zone (Collins et al., 2014; Santosh et al., 2012). This sector forms a part of the S verging back thrust system, which has been described as a crustal scale "flower structure," (Chetty and Bhaskar Rao, 2006a), suggesting transpression analogous to modern collisional belts. Precambrian ophiolitic rocks derived from suprasubduction zone setting were reported from nearby areas. *(T.R.K. Chetty)*

FIGURE 1.57 **Ptygmatic folds of isoclinals, round hinge and box geometries in banded calc-silicate gneiss hosted in a recrystallized marble matrix.** The calc-silicate gneiss was more viscous than the matrix. 25-mm diameter coin for scale. N to Munsiari, India. *(Subhadip Mandal)*

FIGURE 1.58 **Mesoscopic disharmonic folding within the eastern limb of a thrust-related anticline along the NNE-SSW-trending Valnerina Line (Tavarnelli et al., 2001), Umbria-Marche Apennines, Italy.** Folding affected the well-bedded marly-calcareous strata of the Upper Cretaceous-Eocene Scaglia Rossa Formation. The stunning mesoscale polyclinal fold involving the decimeter-scale calcareous strata is characterized by several non-parallel axial planes but subparallel hinge lines. Cataclastic flow of the intervening thin marly layers produces thinned limbs and thickened hinge toward the core of the detachment folds that involve the centimeter-scale calcareous strata. Mesoscopic folds and thrust-related structures from this region of the Northern Apennines are described in Tavarnelli (1996, 1997) and Decandia et al. (2002). Location: Valnerina valley (42.784238° 12.869925°), N to the Piedipaterno village, province of Perugia (Italy). Width of view: 2.1 m. *(Paolo Pace)*

FIGURE 1.59 **Folded quartz vein in dolomitic rock showing different geometries of hinge region.** Central Indian Tectonic Zone. Jabalpur, Madhya Pradesh, India. *(Jyotirmoy Paul)*

FIGURE 1.60 **Boudinaged mafic layer set in quartzofeldspathic paragneiss subsequently folded.** From Mullet Penninsula, Co. Mayo, Ireland. *(Kieran F. Mulchrone, Patrick Meere)*

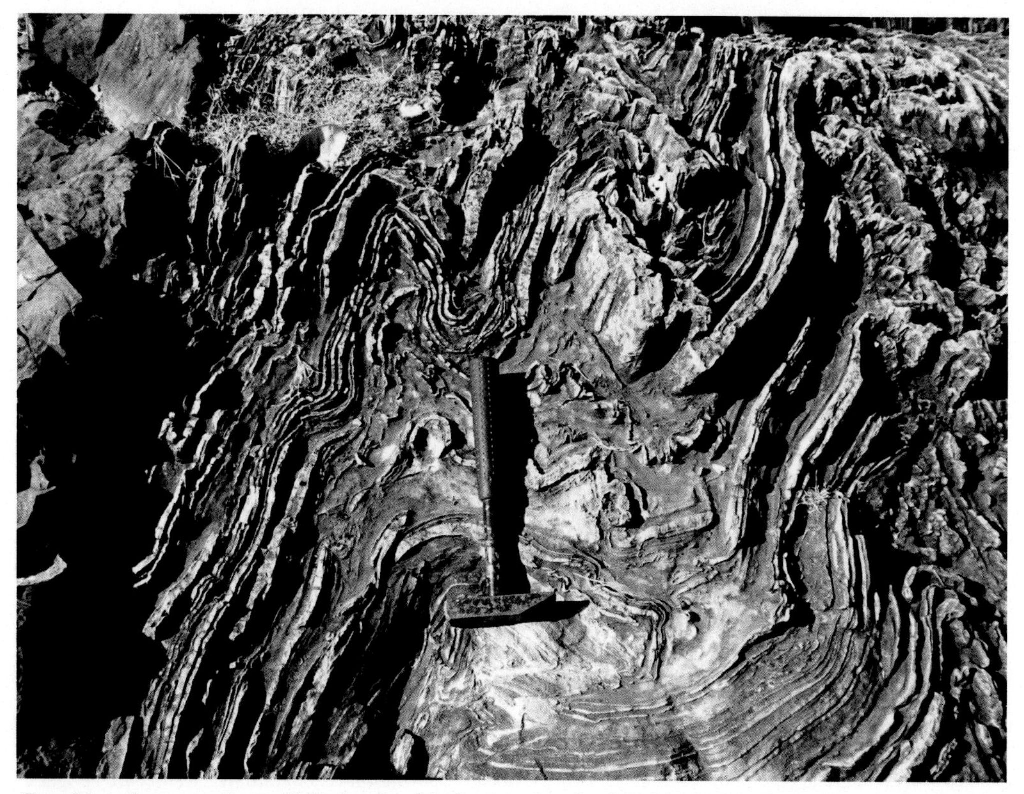

FIGURE 1.61 **Type 3 interference pattern of folds in calc schist in plan view.** South Delhi fold Belt. Taleti village, Ambaji, Gujrat, India. *(Narayan Bose, Soumyajit Mukherjee)*

FIGURE 1.62 **Type 3 interference pattern of folds in amphibolite.** Upper Aravalli Group. Near Majam, Udaipur district, Rajasthan, India. *(Shruthi Narayanan)*

FIGURE 1.63 **Fold interference pattern in the Variscan basement in Northern Sardinia Island (Italy).** Base of the picture: ~130 cm. A thin-layered sedimentary sequence metamorphosed to greenschist facies records three-folding events with orthogonal axial planes and parallel fold axes. Both axis and axial planes are perpendicular to the picture view. The older deformation event produced the asymmetric intrafoliar fold (F1) showed in the left portion of the picture. The fold, developed in white and light-gray quartzites and quartz-rich fine paragneisses is an isoclinal with thinned delaminated limbs and thickened hinge. The F1 axial plane foliation, the main foliation documented in the field, is marked by a compositional layer made of muscovite- and biotite-bearing mica schists (the yellow-brownish portion) and light gray/ochre quartzites. It is deformed by a metric-size upright fold (F2) with an antiform geometry and an interlimb angle ranging from 55° to 80°. The F2 subvertical axial plane is gently folded by the late deformation that produced recumbent open fold (F3) with sub-horizontal axial plane and sub-horizontal axes. The superposition of F2 on F1 fold and those of F3 on F2 folds produce a Type 3 fold interference pattern (Ramsay and Huber, 1987). *(Chiara Frassi)*

FIGURE 1.64 **Doubly plunging fold with prominent axial depression, in meta-sediments of Chaibasa Formation, Tetuldanga, Ghatshila, Jharkhand, India.** *(Rahul Kar)*

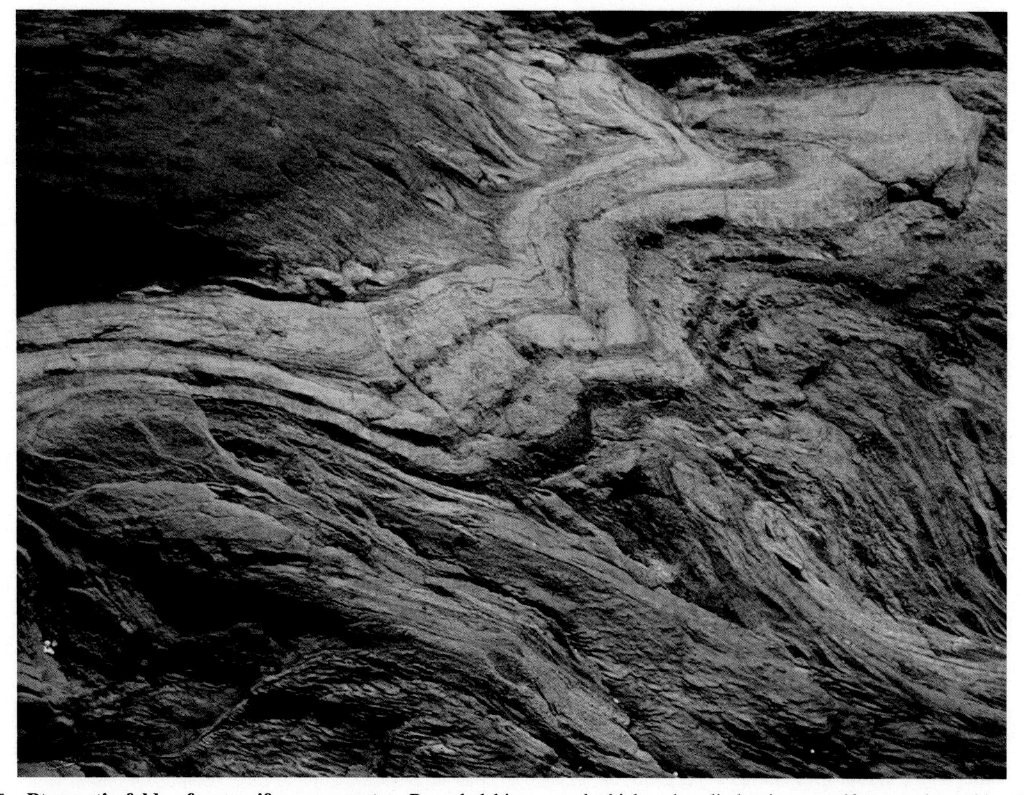

FIGURE 1.65 **Ptygmatic folds of nonuniform geometry.** Rounded hinge much thicker than limbs, in garnetiferous mica schist, Pur, Bhilwara, Rajasthan, India. *(Rahul Kar)*

FIGURE 1.66 **Type III fold interference with a perfect chicken neck structure.** Produced by early tight reclined/inclined fold overprinted by a late set with nearly same axial trend but with vertical axial surface normal to the first set. Neoproterozoic dolomite of Sirohi Group. At Sivagaon, Sirohi district, Rajasthan, India. Outcrop width ~80 cm. *(Shailendra Singh)*

FIGURE 1.67 **Noncylindrical fold with curved axial plane.** An inhomogeneous simple shear component is deciphered. Strain varies parallel to the hinge line and on the plane perpendicular to the hinge. Neoproterozoic banded calc-silicate of Sirohi Group. Khiwandi, Pali district, Rajasthan, India. Marker pen: 15 cm. *(Shailendra Singh)*

FIGURE 1.68 **The field photograph shows a plane noncylindrical fold with a curved hinge line in a calc-silicate mylonitic rock of the Phulad shear zone (Ghosh et al 1999).** The striping lineation (Sengupta and Ghosh 2007) is deformed by the fold. It is perpendicular at the hinge but rotates to make a smaller angle at the limb. The Mylonitic foliation is subvertical with a northeasterly strike and downdip lineation. The width of the photo is 30 cm. Sample location is 25°36′33.1″N, 73°48′45.1″E. *(Sadhana M. Chatterjee, Sudipta Sengupta)*

FIGURE 1.69 **A set of thick and thin quartzofeldspathic layers interlayered in a calcareous matrix show complex folded structures.** Locality is Phulad shear zone, Rajasthan, India. The folds show two generations of disharmonic folding. The axial planes of the dominant folds strike NNE and there is a later set of folds with axial planes at a high angle to the dominant folds. The thin layers show pinch and swell or boudinage structure at the outer arc of the larger folds. At the inner cores of the larger folds, the small folds are very tight or isoclinal. Arrow shows the geographic north direction. The width of the photo is 40 cm. Sample location is 25°36′29.0″N, 73°48′51.0″E. *(Sadhana M. Chatterjee, Sudipta Sengupta)*

FIGURE 1.70 **Conjugate folds in Neoproterozoic marble, Sirohi Group.** Subhorizontal plunge of the upright fold at left. At right, the same fold shows a much steep plunge. Such conjugate folds are produced in ductile shear zones. Revdar, Sirohi district, Rajasthan, India. Diameter of the lens cover: 6cm. *(Shailendra Singh)*

FIGURE 1.71 **Folded quartz vein within meta-greywacke depicting conjugate fold geometry.** Kumbhalgarh Formation, ~85km northwest of Udaipur, Rajasthan, India. *(Swagato Dasgupta)*

FIGURE 1.72 **Complex, disharmonic folding in lacustrine organic rich dolomitic lime mudstone/"oil-shale" deposits of the Eocene Green River Formation, Piceance Creek Basin, Colorado, USA.** Folds represented by close and tight folds. Deformation is situated in a brecciated oil shale bed, over and underlain by undeformed laminated oil shale deposits. These sediments/"contorted breccias" "disrupted" or "streaked-and-blebby" oil shale beds in the literature range from a few centimeters up to eleven meter thick, with flat/horizontal upper and lower boundaries. They generally originate from slumps, slides, debris-flows, and higher-energy turbidity currents and were correlated over several square kilometers in the Piceance Creek Basin (e.g., Dyni and Hawkins, 1981; Johnson, 1981; Tänavsuu-Milkeviciene and Sarg, 2012). The formation of such folds requires nondirectional shear stresses, which might be generated within the mass transport deposits. Alternatively, considering the deposition in a very gentle/flat lake bottom, such shear stresses can be also generated by seismic shaking, and consequently, the occurrence of these deposits indicates synsedimentary tectonic activity and deformation related to seismic shaking. *(Reproduced from Figure 12 of Tőrő et al. (2013)—Balázs Tőrő, Brian R. Pratt, Sudipta Dasgupta)*

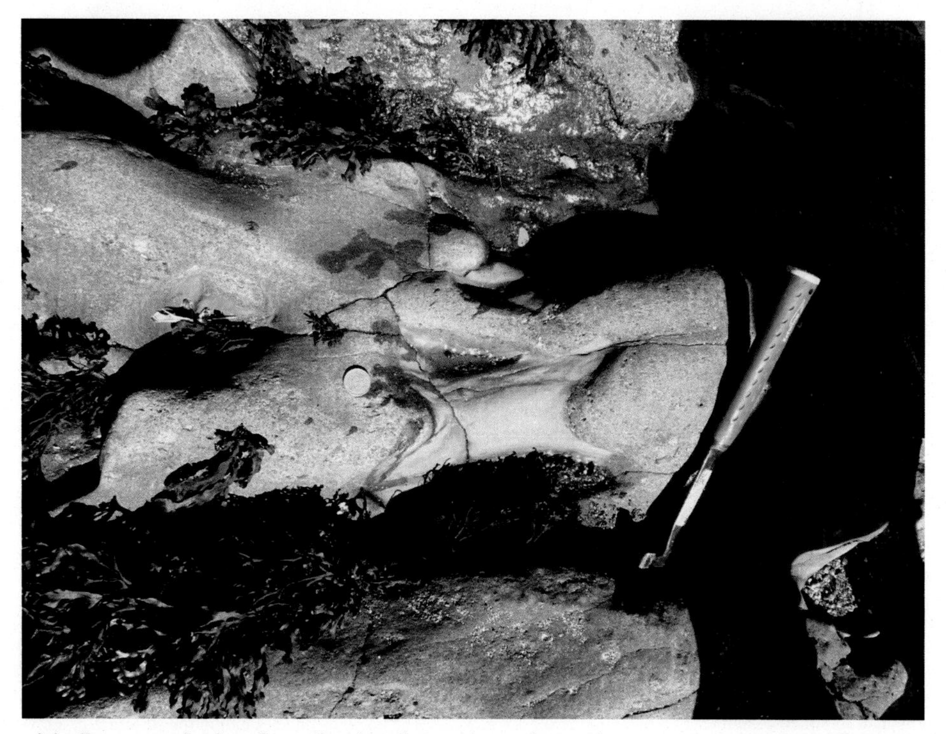

FIGURE 1.73 **An originally rectangular intraformational mudstone rip-up clast set in a coarse red sandstone which has been deformed into a "butterfly shape."** Is this a dome and basin Type-1 superposed folding? This is because the clast is less competent than the surrounding material. From Wine Strand, Ballyferriter, Co. Kerry, Ireland. *(Kieran F. Mulchrone, Patrick Meere)*

FIGURE 1.74 **Elongated dome and basin structures on a subvertical (100/78S) ferruginous quartzite layer developed due to interference of two sets of open folds.** The early folds are subvertical and the later folds are subhorizontal with a gentle plunge toward west. The structure show first mode of superposed buckling where the interlimb angle of the early folds was more than 135° during later folding (Ghosh, 1993, page 340, Ghosh. et al., 1992, Sengupta et al., 2005). On a parallel surface (on the left side), the smaller subhorizontal later folds are more dominant and tighter. There is a set of small wrinkles (lower part of the photo) parallel to the subhorizontal folds. ~4 km from Kemmangundi, Bababudan Hills, Karnataka. The length of the pen in the lower left is 15 cm. GPS Location: N13°33′/29″ and E75°45′27″. *(Sudipta Sengupta, Sadhana M. Chatterjee)*

FIGURE 1.75 **1 to 3 cm-thick, folded anhydrite layers fully encased in rock salt exposed in a solution-mined cavern.** At the level of a few hundred meters depth, this salt mine in the Zechstein 3 sequence in Eastern Germany (ESCO salt mine Bernburg) frequently shows stacks of thin anhydrite layers that folded during salt deformation. The complex structure of folds indicates superposed folding that most likely emerges from complex salt flows. As the mechanical behavior of anhydrite during salt deformation is still poorly understood, the fold morphologies of these layers help estimating the competence contrast between the anhydrite and the rock salt (Schmalholz and Podladchikov, 2001) as well as understanding the structure of anhydrite layers as tracers for intrasalt deformation also on the larger scale (see Van Gent et al., 2011 and Strozyk et al., 2012). Deformation and evolution of salts in a larger scale, such as in salt domes have been modeled by many (e.g., Mukherjee et al., 2010; Mukherjee, 2011b). Photo width ~50 cm. *(Frank Strozyk)*

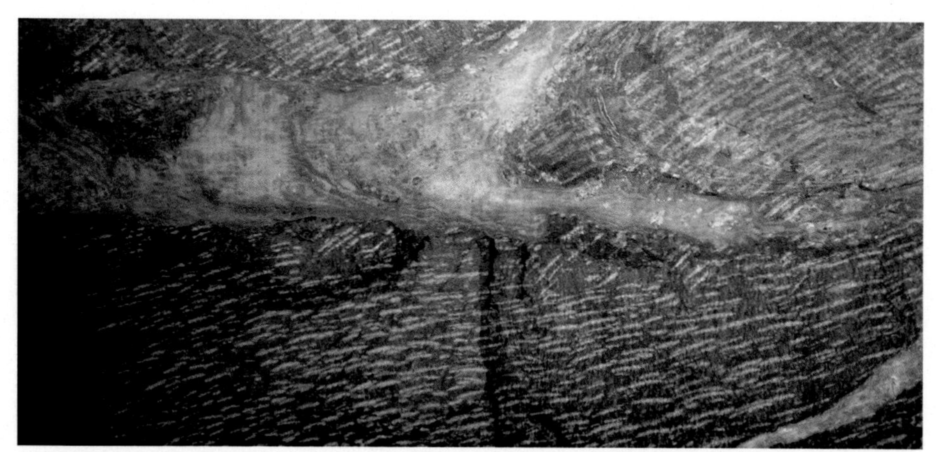

FIGURE 1.76 **Red salt in fractured anhydrite.** Thickness and competence contrast for layered evaporites cause brittle and ductile deformation at various scales. Compared to folding of thin anhydrite layers embedded in salt, thick layers of anhydrite also tend to break and boudinage during salt deformation. From internal part of one of the formed anhydrite blocks from the Zechstein 3 sequence in a salt mine in Eastern Germany, where the Z3 main anhydrite (dark) is several tens of meters thick and show dominantly brittle deformation with an internal fracture system (brighter colors). The fractures are filled with fine-grained halite that shows complicated internal flow structures. The red color of the halite is a typical feature at the anhydrite-halite-contact, and most likely associate with fluid interactions between the partially overpressured anhydrite and the salt (Hemmann, 1972). Such analog examples of brittle evaporite deformation during ductile salt flows help studying the mechanical interaction between the evaporites during deformation (Zulauf et al., 2009). This is important to understand the short- and long-term mechanics and fluid interactions of layered evaporites. These factors govern the dynamics and evolution of salt-basins, which host significant volume of oil and gas, and serve for salt mining as well as storage caverns and waste repository sites. Photo width ~120 cm. *(Frank Strozyk)*

FIGURE 1.77 **Anhydrite diapir in rock salt.** The dark, mushroom-shaped diapir structure consists of Zechstein 3 anhydrite fully encased in halite in a salt mine in Eastern Germany. The anhydrite is layered and shows complex folding inside the diapir, which is of similar shape and internal structure like large salt diapirs. We assume that it developed coevally to the transformation of gypsum to anhydrite (see e.g., Williams-Straud and Paul, 1997), and that the quicksand-like material squeezed into the overlying rock salt. This process might have been responsible for the common observation of spatially thickened anhydrite (anhydrite "klippen": Zeibig and Wenzel, 2000) in the Zechstein Basin. Thin, black clay layers are often associated to the main anhydrite and can be found along the stem of the diapir. Here, the clay was dragged with the gypsum during the formation of the structure and also interacted with the deforming, surrounding rock salt, as indicated by a thin layer of clay tracing a very complicated, overturned structure in the salt left of the diapir stem. Such spectacular features exposed in salt mines help estimating the relative timing and origin of the formation of intrasalt structures, and indicate that some of the structures we observe are not a product of salt tectonics, but formed during shallow burial diagenesis of the evaporites. Width of picture: ~6 m. *(Frank Strozyk)*

FIGURE 1.78 **Structures of Potassium salt.** Accumulations of potassium salts i.e., carnallite in the uppermost evaporite sequences are a common feature in German Zechstein salt mines (ESCO salt mine, Gronau, Eastern Germany). The colorful, reddish to white salts alter with layers of halite and anhydrite, and form very complex structures (Urai, 1987). Larger folds of anhydrite and halite are often filled by internally complicated folds of the K-salts, which squeeze very fast and can be of highly variable thickness. However, the mechanical behavior of these evaporites during active salt deformation is still poorly understood. Same applies for their genesis and the environment of precipitation, as their fast and complex deformation overprints the original position, geometry, and extension of the layers. Finally, their hygroscopic behavior overprints features that actually formed during salt deformation. Analog and numerical models are required to study them. This would be of economic interest. Width of picture: ~3 m. *(Frank Strozyk)*

FIGURE 1.79 **Fractured ptygmatic fold of quartz vein within granite gneiss.** The fold has a thick hinge and narrowing limbs, of 1C geometry. N23°50.715′, E86°57.152′. India. *(Ananya Basu)*

FIGURE 1.80 **Complex folding with near horizontal E-W fold plunges derived from N verging thrust in a W looking sectional view.** Fold structures are defined by cherty bands associated with highly weathered mafic–ultramafic rock assemblages. The outcrop is located along a newly dug Rail cutting section between Namakkal and Mohanur and occurs along the WNW-ESE trending S boundary zone of the Cauvery Suture Zone, India (Chetty and Bhaskar Rao, 2006b). Interestingly, the entire assemblage represents a package of mafic–ultramafic rocks defining the back thrust system of the E-W trending crustal scale "flower structure" (Chetty and Bhsakar Rao, 2006b). Ultra high temperature (UHT) mineral assemblages are also reported in the vicinity (Santosh et al., 2004). These are also interpreted to have been generated through suprasubduction zone setting representing ocean plate stratigraphy (Yellappa et al., 2012; Praveen et al., 2014). *(T.R.K. Chetty)*

FIGURE 1.81 **Type 2 interference pattern of folds in amphibolite.** Upper Aravalli Group. Near Majam, Udaipur district, Rajasthan, India. *(Shruthi Narayanan)*

FIGURE 1.82 **Two-generation folding in lower greenschist facies quartzite of Lower to Middle Proterozoic Aravalli Supergroup.** Near Zawarmala ~25 km S to Udaipur, Rajasthan, India. Limbs of first phase folds crenulated to produce the second phase fold. The first phase fold axis parallels to that of the second fold. *(Eirin Kar)*

FIGURE 1.83 **Type 3 interference pattern or hook-shaped pattern in the high-grade quartzite rock at the contact between the Delhi Supergroup with the Mesoproterozoic basement gneisses.** Near Shrinagar ~15 km from Ajmer toward NE, Rajasthan, India. *(Eirin Kar)*

FIGURE 1.84 **"Gravifossum" structure in trough cross-stratified sandstone Early Pleistocene Lower Morne L'Enfer Members, SW Trinidad.** "Gravifossum" (van Loon, 2009) is a soft-sedimentary deformation structure formed at the irregular interface between two fluids with different densities (unlithified sands with two different fluid pressures here), when one fluid layer accelerates into the other by fluid dynamic Rayleigh–Taylor instability, often aided and accentuated by passage of seismic shock-wave (Richtmyer-Meshkov instability). Montenat et al. (2007) described similar structures as thixotropic wedges, and interpreted as localized liquefaction and sediment collapse features formed by seismic shaking. The sands were deposited by the subaqueous distributary hyperpycnal channels on the outer shelf feeding the mouth-bar lobes of the palaeo-Orinoco river at the shelf-edge. *(Balázs Törő, Sudipta Dasgupta, Luis A. Buatois)*

FIGURE 1.85 **Doubly plunging fold in the Late Paleocene to Eocene sandstone along the Koshalia River section near the Koti area, Himachal Pradesh, India.** *(Eirin Kar)*

FIGURE 1.86 **"Gravifossum" structure in heterolithic sediments of laminated siltstones and sandstones Early Pleistocene Upper Morne L'Enfer Members, SW Trinidad.** "Gravifossum" (van Loon, 2009) is a soft-sedimentary deformation structure formed at the irregular interface between two fluids with different densities (unlithified silt and sand laminae here), when one fluid layer accelerates into the other by fluid dynamic Rayleigh–Taylor instability, often aided and accentuated by passage of seismic shock-wave (Richtmyer-Meshkov instability). Montenat et al. (2007) described similar structures as thixotropic wedges, and interpreted as localized liquefaction and sediment collapse features formed by seismic shaking. The heterolithic units were deposited by the subaqueous distributary hyperpycnal channels on the outer shelf feeding the mouth-bar lobes of the palaeo-Orinoco river at the shelf-edge. *(Balázs Törő, Sudipta Dasgupta, Luis A. Buatois)*

FIGURE 1.87 **Feldspathic layer in mafic mineral-rich gneiss.** Are we looking at a section through a sheath fold? *(Kieran F. Mulchrone, Patrick Meere)*

REFERENCES

Acharyya, S.K., Saha, P., 2008. Geological setting of the Siang Dome located at the Eastern Himalayan Syntaxis. Himalayan Journal of Science 5, 16–17.

Alsop, G.I., Holdsworth, R.E., 2004. Shear zone folds: records of flow perturbation or structural inheritance? In: Alsop, G.I., Holdsworth, R.E., McCaffey, K.J.W., Hand, M. (Eds.), Flow Processes in Faults and Shear Zones, vol. 224. Geol Soc London, Spec Publ, pp. 177–199.

Anderson, J.R., Payne, J.L., Kelsey, D.E., Hand, M., Collins, A.S., Santosh, M., 2012. High-pressure granulites at the dawn of the Proterozoic. Geology 40, 431–434.

Bell, T.H., 2010. Deformation partitioning, foliation successions and their significance for orogenesis: hiding lengthy deformation histories in mylonites. In: Law, R.D., Butler, R.W.H., Holdsworth, R.E., Krabbendam, M., Strachan, R.A. (Eds.), Continental Tectonics and Mountain Building: The Legacy of Peach and Horne, vol. 335. Geol Soc London, Spec Publ, pp. 275–292.

Bell, T.H., Rudenach, M.J., 1983. Sequential porphyroblast growth and crenulation cleavage development during progressive deformation. Tectonophysics 92, 171–194.

Bikramaditya Singh, R.K., Gururajan, N.S., 2011. Microstructures in quartz and feldspars of the Bomdila gneiss from western Arunachal Himalaya, India: Implications for geotectonic evolution of the Bomdila mylonitic zone. Journal of Asian Earth Sciences 42, 1163–1178.

Calamita, F., Pizzi, A., Ridolfi, M., Rusciadelli, G., Scisciani, V., 1998. Il buttressing delle faglie normali sinsedimentarie pre-thrusting sulla strutturazione neogenica della catena appenninica: l'esempio della M.gna dei Fiori (Appennino Centrale esterno). Bollettino Della Società Geologica Italiana 117, 725–745.

Calamita, F., Ben M'Barek, M., Di Vincenzo, M., Pelorosso, M., 2004. The Pliocene thrust system of the Gran Sasso salient (Central Apennines, Italy). In: Pasquarè, G., Venturini, C. (Eds.), Mapping Geology in Italy. S.EL.CA, Florence, Italy, pp. 227–234.

Calamita, F., Esestime, P., Paltrinieri, W., Scisciani, V., Tavarnelli, E., 2009. Structural inheritance of pre- and syn- orogenic normal faults on the arcuate geometry of Pliocene-quaternary thrusts: examples from the Central and Southern Apennine Chain. Italian Journal of Geosciences 128, 381–394.

Calamita, F., Satolli, S., Scisciani, V., Esestime, P., Pace, P., 2011. Contrasting styles of fault reactivation in curved orogenic belts: examples from the Central Apennines (Italy). Geological Society of America Bulletin 123, 1097–1111.

Calamita, F., Pace, P., Satolli, S., 2012a. Coexistence of fault-propagation and fault-bend folding in curve-shaped foreland fold-and-thrust belts: examples from the Northern Apennines (Italy). Terra Nova 24, 396–406.

Carreras, J., Druguet, E., Griera, A., 2005. Shear zone-related folds. Journal of Structural Geology 27, 1229–1251.

Chetty, T.R.K., 1996. Proterozoic shear zones in southern granulite terrain, India. In: Santosh, M., Yoshida, M. (Eds.), Gondwana Research Group Memoir-3: The Archaean and Proterozoic Terrains in Southern India within East Gondwana, pp. 77–89.

Chetty, T.R.K., Bhaskar Rao, Y.J., 2006a. The Cauvery Shear Zone, southern Granulite Terrain, India: a crustal-scale flower structure. Gondwana Research 10, 77–85.

Chetty, T.R.K., Bhaskar Rao, Y.J., 2006b. Constrictive deformation in transpressional regime: field evidence from the Cauvery Shear Zone, southern Granulite Terrain, India. Journal of Structural Geology 28, 713–720.

Collins, A.S., Clark, C., Plavsa, D., 2014. Peninsular India in Gondwana: The tectonothermal evolution of the Southern Granulite terrain and its Gondwanan counterparts. Gondwana Research 25, 190–203.

Crowe, W.A., Nash, C.R., Harris, L.B., Leeming, P.M., Rankin, L.R., 2003. The geology of the Rengali Province: implications for the tectonic development of northern Orissa, India. Journal of Asian Earth Sciences 21, 697–710.

Decandia, F.A., Tavarnelli, E., Alberti, M., 2002. Pressure-solution fabrics and their overprinting relationships within a minor fold train of the Umbria-Marche Apennines, Italy. Bollettino Della Società Geologica Italiana 1, 687–694.

Deng, H., Zhang, C., Koyi, H.A., 2013. Identifying the characteristic signatures of fold-accommodation faults. Journal of Structural Geology 56, 1–19.

Drury, S.A., Holt, R.W., 1980. The tectonic framework of the south Indian Craton, a reconnaissance involving Landsat imagery. Tectonophysics 65, T1–T15.

Dyni, J.R., Hawkins, J.E., 1981. Lacustrine turbidites in the Green river formation, northwestern Colorado. Geology 9, 235–238.

Ez, V., 2000. When shearing is a cause of folding. Earth-Science Reviews 51, 155–172.

Fagereng, A., Sibson, R.H., 2010. Melange rheology and seismic style. Geology 38, 751–754.

Ghosh, S.K., Mandal, N., Khan, D., Deb, S.K., 1992. Modes of superposed buckling in single layers controlled by intial tightness of early folds. Journal of Structural Geology 14, 381–394.

Ghosh, S.K., 1993. Structural Geology: Fundamentals and Modern Developments. Pergamon Press, Oxford.

Ghosh, S.K., Hazra, S., Sengupta, S., 1999. Planar, non-planar and refolded sheath folds in Phulad shear zone, Rajasthan, India. Journal of Structural Geology 21, 1715–1729.

Ghosh, J.G., de Wit, M.J., Zartman, R.E., 2004. Age and tectonic evolution of Neoproterozoic ductile shear zones in the Southern Granulite Terrain of India, with implications for Gondwana Studies. Tectonics 23, TC3006.

Godin, L., Yakymchuk, C., Harris, L.B., 2011. Himalayan hinterland-verging superstructure folds related to foreland-directed infrastructure ductile flow: Insights from centrifuge analogue modeling. Journal of Structural Geology 33, 329–342.

Gopalakrishnan, K., 1994. An overview of Southern Granulite Terrain, India—Constraints in Reconstruction of Precambrian Assembly of Gondwanaland, vol. 2, Oxford and IBH Publication, Gondwana Nine, pp.1003–1026.

Grasemann, B., Martel, S., Passchier, C., 2005. Reverse and normal drag along a fault. Journal of Structural Geology 27, 999–1010.

Harris, L.B., 2003. Folding in high-grade rocks due to back-rotation between shear zones. Journal of Structural Geology 25, 223–240.

Harris, L.B., Godin, L., Yakymchuk, C., 2012a. Regional shortening followed by channel flow induced collapse: a new mechanism for—dome and keel geometries in Neoarchaean granite-greenstone terrains. Precambrian Research 212–213, 139–154.

Harris, L.B., Koyi, H.A., Fossen, H., 2002. Mechanisms for folding of high-grade rocks in extensional tectonic settings. Earth-Science Reviews 59, 163–210.

Harris, L.B., Yakymchuk, C., Godin, L., 2012b. Implications of centrifuge simulations of channel flow for opening out or destruction of folds. Tectonophysics 526–529, 67–87.

Hemmann, M., 1972. Ausbildung und Genese des Leinesteinsalzes und des Hauptanhydrits (Zechstein 3) im Ostteil des Subherzynen Beckens. Ber, deutsch. Ges. Geol. Wiss. B. Min. Lagerstättenf 16, 307–411.

Hibbard, M.J., 1995. Petrography to petrogenesis. Chapter 21 "Metamorphic Rocks". Prentice-Hall, New Jersey. 587 p.

Hobbs, B.E., Means, W.D., Williams, P.F, 1976. An Outline of Structural Geology. John Wiley & Sons, New York.

Hudleston, P.J., Lan, L., 1993. Information from fold shapes. Journal of Structural Geology 15, 253–264.

Hudleston, P.J., Treagus, S.H., 2010. Information from folds. a review. Journal of Structural Geology 32, 2042–2071.

Johnson, R.C., 1981. Preliminary Geologic Map of the Desert Gulch Quadrangle, Garfield County, Colorado. U.S. Geological Survey Miscellaneous Field Investigations Map MF-1328, scale 1:24,000.

Khanal, S., Robinson, D.M., Mandal, S., Simkhada, P., 2014. Structural, geochronological and geochemical evidence for two distinct thrust sheets in the 'Main Central thrust zone', the Main Central thrust and Ramgarh–Munsiari thrust: implications for upper crustal shortening in central Nepal. In: Mukherjee, S., Carosi, R., van der Beek, P.A., Mukherjee, B.K., Robinson, D.M. (Eds.), Tectonics of the Himalaya. Special Publications, Geological Society, London, 412. http://dx.doi.org/10.1144/SP412.2.

Kronenberg, A.K., Kirby, S.H., Pinkstone, J., 1990. Basal slip and mechanical anisotropy of biotite. Journal of Geophysical Research 95, 257–278.

Lin, A., 1997. Ductile deformation of biotite in foliated cataclasite, Iida-Matsukawa fault, central Japan. Journal of Asian Earth Sciences 15, 407–411.

Llorens, M.-G., Bons, P.D., Griera, A., Gomez-Rivas, E., Evans, L.A., 2013. Single layer folding in simple shear. Journal of Structural Geology 50, 209–220.

Mortimer, N., 2004. New Zealand's geological foundations. Gondwana Research 7, 261–272.

Mandal, N., Samanta, S.K., Chakraborty, C., 2004. Problem of folding in ductile shear zones: a theoretical and experimental investigations. Journal of Structural Geology 26, 475–489.

Marco, S., Agnon, A., 1995. Prehistoric earthquake deformations near Masada, Dead Sea graben. Geology 23, 695–698.

Mitra, S., 2002. Fold-accommodation faults. AAPG Bulletin 86, 671–693.

McClay, K.R., Insley, M.W., 1986. Duplex structures in the lewis thrust sheet. Crowsnest Pass, Rocky mountains, Alberta, Canada. Journal of Structural Geology 8, 911–922.

Montenat, C., Barrier, P., Ott d'Estevou, P., Hibsch, C., 2007. Seismites: an attempt at critical analysis and classification. Sedimentary Geology 196, 5–30.

Mukherjee S., 2007. Geodynamics, Deformation and Mathematical Analysis of Metamorphic Belts of the NW Himalaya. (Unpublished Ph.D. thesis). Indian Institute of Technology Roorkee. pp. 1–267.

Mukherjee, S., 2010. Microstructures of the Zanskar shear zone. Earth Science India 3, 9–27.

Mukherjee, S., 2011a. Flanking microstructures of the Zanskar shear zone, west Indian Himalaya. YES Bulletin 1, 21–29.

Mukherjee, S., 2011b. Estimating the viscosity of rock Bodies-a comparison between the Hormuz and the Namakdan salt diapirs in the Persian Gulf, and the Tso Morari gneiss dome in the Himalaya. Journal of Indian Geophysical Union 15, 161–170.

Mukherjee, S., 2013. Deformation Microstructures in Rocks. Springer.

Mukherjee, S., 2014a. Atlas of Shear-Zone Structures in Meso-scale. Springer.

Mukherjee, S., 2014b. Review of the flanking structures in meso- and micro-scales. Geological Magazine 151, 957–974.

Mukherjee, S., Koyi, H.A., 2009. Flanking microstructures. Geological Magazine 146, 517–526.

Mukherjee, S., Koyi, H.A., 2010. Higher Himalayan Shear Zone, Zanskar Indian Himalaya- microstructural studies and extrusion mechanism by combination of simple shear and channel flow. International Journal of Earth Sciences 99, 1083–1110.

Mukherjee, S., Punekar, J.N., Mahadani, T., Mukherjee, R. Intrafolial folds- review & examples from the western Indian Higher Himalaya. In: Mukherjee S., Mulchrone K.F. (Eds.), Ductile Shear Zones: From Micro- to Macro-scales. Wiley-Blackwell, in press.

Mukherjee, S., Talbot, C.J., Koyi, H.A., 2010. Viscosity estimates of salt in the Hormuz and Namakdan salt diapirs, Persian Gulf. Geological Magazine 147, 497–507.

Mukhopadhyay, B., Bose, M.K., 1994. Transitional granulite-eclogite facies metamorphism of basic supracrustal rocks in a shear zone complex in the Precambrian shield of south India. Mineralogical Magazine 58, 97–118.

Nelson, K.D., 1982. A suggestion for the origin of mesoscopic fabric in accretionary mélange, based on features observed in the Chrystalls Beach Complex, South Island, New Zealand. Geological Society of America Bulletin 93, 625–634.

Osanai, Y., Nogi, Y., Baba, S., Nakano, N., Adachi, T., Hokada, T., Toyoshima, T., Owada, M., 2013. Geologic evolution of the Sør Rondane Mountains, East Antarctica: collision tectonics proposed based on metamorphic processes and magnetic anomalies. Precambrian Research 234, 8–29.

Passchier, C.W., Trouw, R.A.J., 1996. Microtectonics. Springer-Verlag. p. 289.

Passchier, C., 2001. Flanking structures. Journal of Structural Geology 23, 951–9962.

Passchier, C., Trouw, R., 2005. Microtectonics. Chapter 4 "Foliations, Lineations and Lattice Prefered Orientation". Springer, Berlin. 336 p.

Passchier, C., Trouw, R., 2005. Microtectonics. Chapter 7 "Porphyroblasts and Reaction Rims". Springer, Berlin. p. 336.

Praveen, M.N., Santosh, M., Yang, Q.Y., Zhang, Z.C., Huang, H., Singanenjam, S., Sajinkumar, K.S., 2014. Zircon U–Pb geochronology and Hf isotope of felsic volcanics from Attappadi, India: Implications for Neoarchean convergent margin tectonics. Gondwana Research 26, 907–924.

Ramsay, J.G., 1967. Folding and Fracturing of Rocks. McGraw Hill, New York. pp. 117, 352, 390, 413, 414.

Ramsay, J.G., Huber, M.I., 1987. The Techniques of Modern Structural Geology. Folds and Fractures, vol. 2, Academic Press, London, Orlando, San Diego, New York, Austin, Boston, Sydney, Tokyo, Toronto. 391 pp.

Rodríguez-Pascua, M.A., Calvo, J.P., De Vicente, G., Gómez-Gras, D., 2000. Softsediment deformation structures interpreted as seismites in lacustrine sediments of the Prebetic Zone, SE Spain, and their potential use as indicators of earthquake magnitudes during the Late Miocene. Sedimentary Geology 135, p. 117–135.

Saito, Y., Tsunogae, T., Santosh, M., Chetty, T.R.K., Horie, K., 2011. Neoarchean high pressure metamorphism from the northern margin of the Palghat–Cauvery suture zone, southern India: petrology and zircon SHRIMP geochronology. Journal of Asian Earth Sciences 40, 268–285.

Santosh, M., Maruyama, S., Sato, K., 2009. Anatomy of a Cambrian suture in Gondwana: Pacific-type orogeny in southern India? Gondwana Research 16, 321–341.

Santosh, M., Tsunogae, T., Koshimoto, S., 2004. First report of sapphirine-bearing rocks from the Palghat-Cauvery shear zone system, southern India. Gondwana Research 7, 620–626.

Santosh, M., Xiao, W.J., Tsunogae, T., Chetty, T.R.K., Yellappa, T., 2012. The Neoproterozoic subduction complex in southern India: SIMS zircon U–Pb ages and implications for Gondwana assembly. Precambrian Research 190–208.

Satolli, S., Speranza, F., Calamita, F., 2005. Paleomagnetism of the Gran Sasso range salient (central Apennines, Italy): pattern of orogenic rotations due to translation of a massive carbonate indenter. Tectonics 24, TC4019.

Schmalholz, S.M., Podladchikov, Y.Y., 2001. Strain and competence contrast estimation from fold shape. Tectonophysics 340, 195–213.

Scisciani, V., Tavarnelli, E., Calamita, F., 2002. The interaction of extensional and contractional deformations in the outer zones of the Central Apennines, Italy. Journal of Structural Geology 24, 1647–1658.

Séguret M., 1972. Etude tectonique des nappes et séries décollées de la partie centrale du versant sud des Pyrénées (Ph.D. thesis). Caractère synsédimentaire, rôle de la compression et de la gravité. University of Montpellier.

Sengupta, S., Ghosh, S.K., 2004. Analysis of transpressional deformation from geometrical evolution of mesoscopic structures from Phulad Shear Zone, Rajasthan, India. Journal of Structural Geology 26, 1961–1976.

Sengupta, S., Ghosh, S.K., Deb, S.K., Khan, D., 2005. Opening and closing of folds in superposed deformations. Journal of Structural Geology 27, 1282–1299.

Sengupta, S., Ghosh, S.K., 2007. Origin of striping lineation and transposition of linear structures in shear zones. Journal of Structural Geology 29, 273–287.

Strozyk, F., van Gent, H., Urai, J.L., Kukla, P.A., 2012. 3D seismic study of complex intra-salt deformation: an example from the Zechstein 3 stringer in the western Dutch offshore. In: Alsop, G.I., Archer, S.G., Hartley, A.J., Grant, N.T., Hodgkinson, R. (Eds.), Salt Tectonics, Sediments and Prospectivity, vol. 363, Geological Society, London, pp. 489–501. Special Publications.

Tänavsuu-Milkeviciene, K., Sarg, F.J., 2012. Evolution of an organic-rich lake basin – stratigraphy, climate and tectonics: Piceance Creek basin, Eocene Green River Formation. Sedimentology 59, 1735–1768.

Tavarnelli, E., 1996. The effects of pre-existing normal faults on thrust ramp development: an example from the northern Apennines, Italy. Geologische Rundschau 85, 363–371.

Tavarnelli, E., 1997. Structural evolution of a foreland fold-and-thrust belt: the Umbria-Marche Apennines, Italy. Journal of Structural Geology 19, 523–534.

Tavarnelli, E., Decandia, F.A., Renda, P., Tramutoli, M., Gueguen, E., Alberti, M., 2001. Repeated reactivation in the Apennine-Maghrebide system, Italy: a possible example of fault-zone weakening? In: Holdsworth, R.E., Strachan, R.A., Magloughlin, J.F., Knipe, R.J. (Eds.), The Nature and Tectonic Significance of Fault Zone Weakening, vol. 186, Geological Society, London, pp. 273–286. Special Publication.

Ternet, Y., Baudin, T., Laumonier, B., Barnolas, A., Gil-Peña, I., Martín-Alfageme, S., 2008. Mapa Geológico de los Pirineos a E. 1: 400.000. IGME–BRGM. Madrid-Orleans.

Törő, B., Pratt, B.R., Renaut, R.W., 2013. Seismically Induced Soft-sediment Deformation Structures in the Eocene Lacustrine Green River Formation (Wyoming, Utah, Colorado, USA) – a Preliminary Study. GeoConvention2013: Integration, Calgary (Poster abstract).

Toyoshima, T., Owada, M., Shiraishi, K., 1995. Structural evolution of metamorphic and intrusive rocks from the central part of the Sør Rondane Mountains, East Antarctica. Proceedings of the NIPR Symposium on Antarctic Geosciences 8, 75–97.

Toyoshima, T., Osanai, Y., Baba, S., Hokada, T., Nakano, N., Adachi, T., Otsubo, M., Ishikawa, M., Nogi, Y., 2013. Sinistral transpressional and extensional tectonics in Dronning Maud Land, East Antarctica, including the Sør Rondane Mountains. Precambrian Research 234, 30–46.

Urai, J.L., 1987. Development of microstructure during deformation of carnallite and bischofite in transmitted light. Tectonophysics 135, 251–263.

Van Gent, H., Urai, J.L., de Keijzher, M., 2011. The internal geometry of salt structures – a first look using 3D seismic data from the Zechstein of the Netherlands. Journal of Structural Geology 33, 292–311.

Van Loon, A.J., 2009. Soft-sediment deformation structures in siliciclastic sediments: an overview. Geologos 15, 3–55.

Van Loon, A.J., Brodzikowski, K., Gotowała, R., 1984. Structural analysis of kink bands in unconsolidated sands. Tectonophysics 104, 351–374.

Wang, S., Mo, Y., Phillips, R.J., Wang, C., 2014. Karakoram fault activity defined by temporal constraints on the Ayi Shan detachment, SW Tibet. International Geology Review 56, 15–28.

Williams, M.L., Scheltema, K.E., Jercinovic, M.J., 2001. High-resolution compositional mapping of matrix phases: implications for mass transfer during crenulation cleavage development in the Moretown Formation, western Massachusetts. Journal of Strucrural Geology 23, 923–939.

Williams-Straud, S.C., Paul, J., 1997. Initiation and growth of gypsum piercement structures in the Zechstein Basin. Journal of Strucrural Geology 19, 897–907.

Yellappa, T., Chetty, T.R.K., Tsunogae, T., Santosh, M., 2010. The Manamedu Complex: geochemical constraints on Neoproterozoic suprasubduction zone ophiolite formation within the Gondwana suture in southern India. Journal of Geodynamics 50, 268–285.

Yellappa, T., Santosh, M., Chetty, T.R.K., Kwon, S., Park, C., Nagesh, P., Mohanty, D.P., Venkatasivappa, V., 2012. A Neoarchean dismembered ophiolite complex from southern India: geochemical and geochronological constraints on its suprasubduction origin. Gondwana Research 21, 246–265.

Yin, A., 2006. Cenozoic tectonic evolution of the Himalayan orogen asconstrained by along-strike variation of structural geometry, exhumation history, and foreland sedimentation. Earth-Science Reviews 76, 1–131.

Zeibig, S., Wendzel, J., 2000. Exploration of anhydrite wall and klippe structure for an optimized extraction of rock salt in the K+S salt mine Bernburg (Northern Germany). World Salt Symposium, vol. 1, Elsevier, Amsterdam, pp. 193–198.

Zulauf, G., Zulauf, J., Bornemann, O., Kihm, N., Peinl, M., Zanella, F., 2009. Experimental deformation of a single-layer anhydrite in halite matrix under bulk constriction. Part 1: geometric and kinematic aspects. Journal of Structural Geology 31, 460.

Chapter 2

Ductile Shear Zones

KEYWORDS
C-plane; Couette flow; Ductile shear zone; Flanking structure; Mineral fish; Primary shear (C) plane; Pure shear; Shear; Simple shear; Synthetic secondary shear plane.

"Tabular or sheetlike, planar or curviplanar zones in which rocks are more highly strained than rocks adjacent to the zone" are called ductile shear zones (Davis et al., 2012; also see Mukherjee and Biswas, 2014; in press). Identification and study of ductile shear zones (Figures 2.1–2.55) are important since major plate boundaries are defined by such shear zones (Regenauer-Lieb and Yuen, 2003). We need to study such zones since along them partially molten rocks can flow (review by Clark et al., 2011). Secondly, viscous dissipation related to such zones has been investigated (Nabelek et al., 2011). No slip boundary condition was assumed classically to explain kinematics of ductile shear zones (Ramsay 1980; Mukherjee 2012; etc). However, recently, slip boundary condition is more recognized (Frehner et al., 2011; Mulchrone and Mukherjee, submitted). The ductile shear sense/ sense of movement from such zones can be deciphered mainly from asymmetric sigmoid, parallelogram and lenticular clasts and intrafolial folds (Lister and Snoke, 1984; ten Grotenhuis et al., 2003; Mukherjee, 2011a,b, 2013a,b,c, 2014a,b,c; Bhadra and Gupta, in press). See Passchier and Trouw (2005) for review on ductile shear zones, and Mukherjee and Mulchrone (2013) and Mulchrone and Mukherjee (in press) for shear heat pattern in these zones. In addition to such shear sense indicators, this chapter also presents near symmetric clasts that form possibly within shear zones but that do not give the shear sense.

FIGURE 2.1 **Domino style normal faults with strike slip component, dipping 50° toward N60°–64°E (rake 70°W).** Upper Pliocene–Lower Pleistocene polymictic volcanic breccias. A transtension component with western extension, is likely due to the parallelism of the Cordillera with the NW-trending Middle America trench located ~160 km to the W. The Cordillera de Tilarán is an andesitic extinct volcanic range, a paleoarc genetically associated with the Costa Rica subduction zone. There most of the ongoing and recent maximum and minimum horizontal stress are generated. Campos de Oro Arriba (10°22′4.09″N – 84°54′57.77″ W), Guanacaste Province, Costa Rica. See Dabrowski and Graseman (2014) as a latest paper on domino type structure. *(Guillermo Alvarado Induni)*

Atlas of Structural Geology. http://dx.doi.org/10.1016/B978-0-12-420152-1.00002-8

FIGURE 2.2 **Extensional shear bands.** Limestone breccia with reddish matrix consisting of marl and calcarenite is a very characteristic stratigraphic horizon of the Julian Alps, marking the onset of flysch sedimentation in front of the rising Alpine orogen in Late Cretaceous (e.g., Miklavič and Rožič, 2008). Here, subhorizontal bedding planes are displaced by numerous normal faults and shear planes which dip toward left. The outcrop is positioned immediately left of a kilometer-scale normal fault with similar dip and orientation. Fault planes with up to several meters of offset are clearly brittle in character (see e.g., a large dissected boulder in the upper-central part of the picture), but prevailing are numerous discontinuous shear bands, which offset and curve the bedding sigmoidally. While such shear bands commonly develop in higher-grade rocks, they are rarely observed in unmetamorphosed sediments. Their development may have been facilitated by highly heterogeneous composition of the rock, where large rigid clasts localized deformation within fine-grained matrix. This outcrop is the only known example of such deformation in the Julian Alps. Rock type: mud-supported blocky breccia, Upper Flyschoid Formation, Late Cretaceous. Location: Kozjak gorge, Julian Alps, Slovenia. Coordinates: 46°15′50″N, 13°36′30″E. *(Marko Vrabec)*

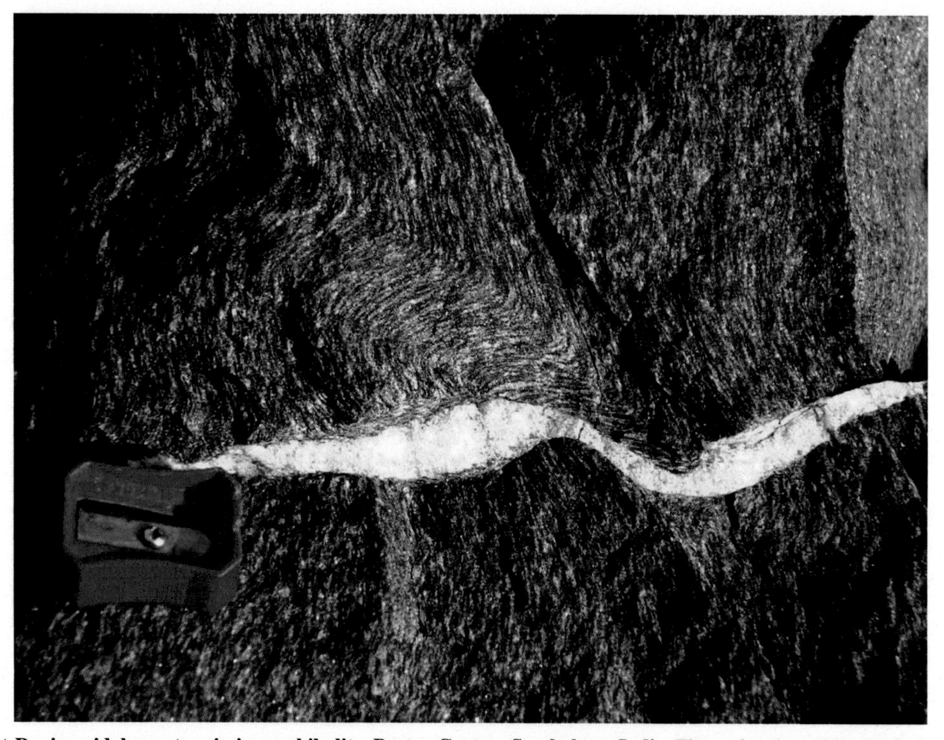

FIGURE 2.3 **Post-D_3 sigmoidal quartz vein in amphibolite, Bastar Craton, Sambalpur, India.** The regional amphibolite facies foliation (S_3), represented by fine segregations of hornblende and plagioclase-rich layers, is crosscut by a post-D_3 quartz vein. S_3 was crenulated, and the vein was folded during subsequent D_4 and D_5 deformation. Top-to-right shear. On National Highway-6, near Padiabahal, Sambalpur district, Odisha, India. *(Arindam Dutta, Saibal Gupta, M.K. Panigrahi)*

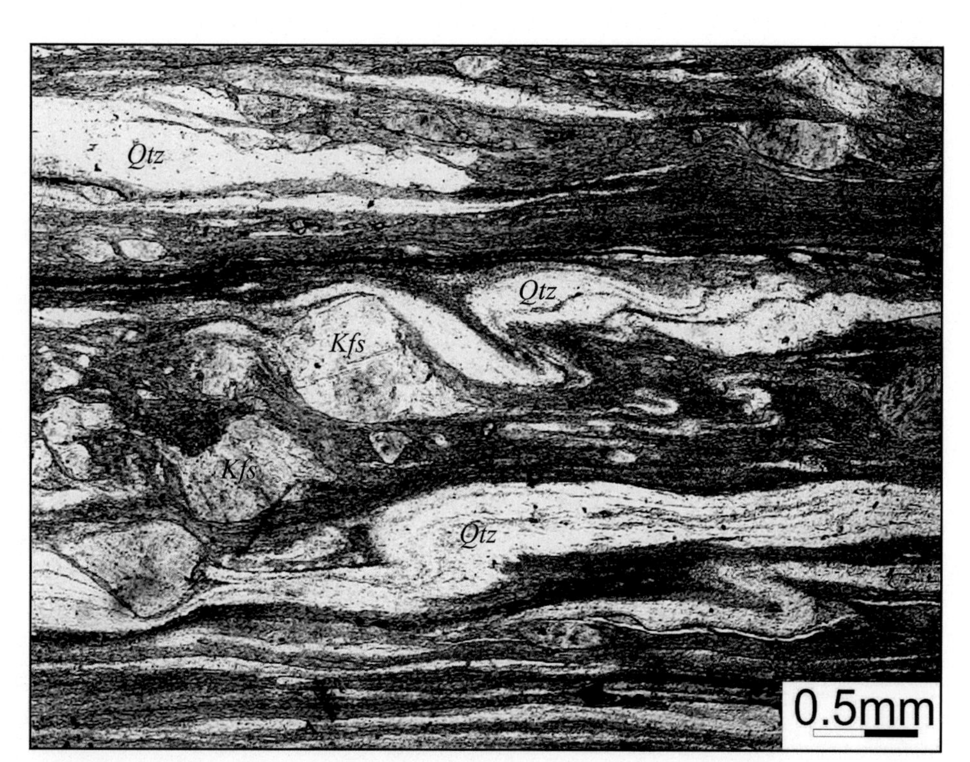

FIGURE 2.4 **Photomicrograph under the plane polarized light of Paleoproterozoic S-type Lesser Himalayan Granitoids from Kameng valley of Western Arunachal Himalaya showing asymmetric microfolds defined by quartz (Qtz) with pressure shadow around k-feldspar (kfs) porphyroclast.** Quartz shows intrafolial folding (Vernon, 2004; Bikramaditya Singh, 2010; also see Mukherjee, 2007, 2010a,b; Mukherjee and Koyi, 2010a,b; Mukherjee, 2013a,b; Mukherjee et al., in press). GPS Location: 27°07′47.2″N; 92°32′53.0″E. *(Bikramaditya Singh)*

FIGURE 2.5 **Ductile sheared quartz vein shows a top-to-right shear.** The sigmoid-shaped vein indicates the ductile shear S-fabric. *(Anupam Samanta)*

FIGURE 2.6 **Foliation parallel train of boudins in schistose quartzite.** The large sigmoid quartz boudin, in the center, is necked at the margins. Top-to-right shear indicated. A symmetric lenticular/rhombic boudin in the same train: no shear sense indicated. Photographed perpendicular to foliation. 14-mm pen for scale. Berinag Quartzite, Sutlej Valley, Himachal Pradesh, NW India. *(Subhadip Mandal)*

FIGURE 2.7 **A relatively large feldspar σ-type porphyroclast.** Top-to-right shear. The outcrop parallels the XZ section. The rock is a metagranite, part of mylonitic zone, which is formed along the contact between the Rila-Rhodope batholith and the surrounding metamorphic "frame." West Rhodope Mountains—Southwest Bulgaria, near the village of Kovachevitsa. *(Svetoslav Bontchev)*

FIGURE 2.8 **A top-to-the left sheared sigmoidal quartz lens in the garnet bearing schist.** Location: Unit I, Greater Himalaya, Marsyangdi section, Central Nepal Himalaya. *(Subodha Khanal)*

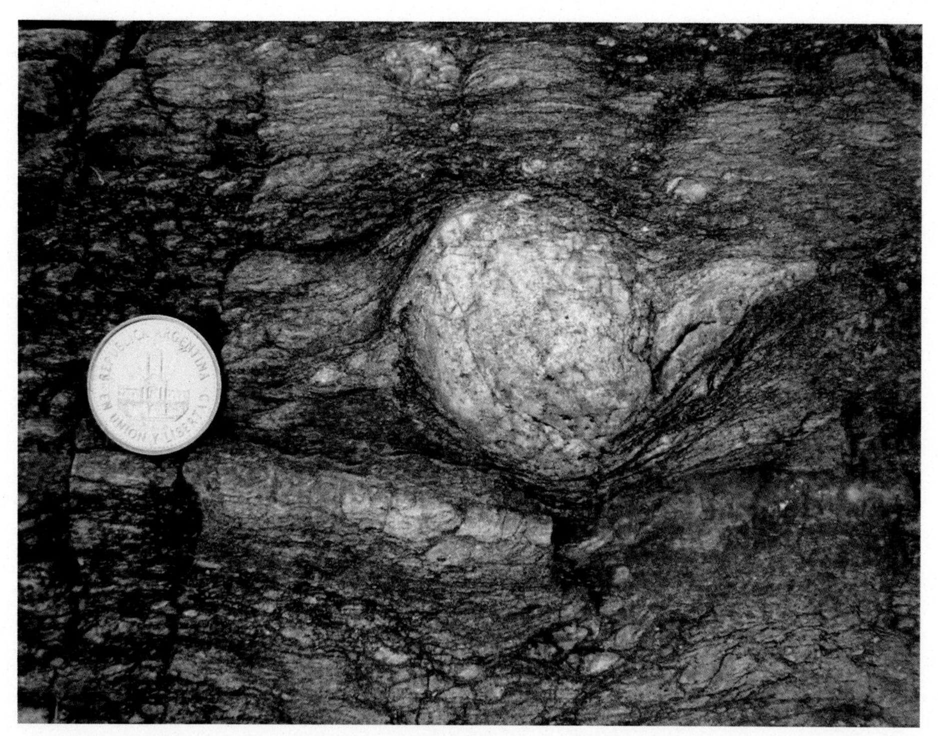

FIGURE 2.9 **Mylonite with a big delta mantled porphyroclast of K-feldspar in a matrix of quartz, mica and feldspar.** Mylonite is derived from a porphyritic granite characterized by large K-feldspar phenocrysts (up to 20 cm long). Medium-grade conditions can be inferred from the recrystallization of K-feldspar along the rims of porphyroclast. Shear sense is dextral as indicated by the stair stepping to the right across the delta structure. Section subparallel to the lineation and normal to the foliation. Width of view 154 mm. Location: Sierra de Velasco, La Rioja Province, NW Argentina; GPS point (WGS84): S 28°36′26.2″ - W 67°10′38.5″; Rock type: Mylonite; Formation name: Ortogneiss Antinaco (TIPA Shear Zone); Age (relative): Post-Lower Ordovician / Pre-Carboniferous; Facies/grade: medium grade. *(Mariano A. Larrovere)*

FIGURE 2.10 **Quartzite veinlet as a crosscutting element within host calc schist rock.** Dragging effect of foliated layer near the quartzite also found at its both margins. Competency contrast probably played significant role in generating the sigmoid shape of the quartzite body. Top-to-left ductile shear. Taleti Village area: (Latitude: 24^0 21.792′N Longitude: 72^0 54.640′E), near Ambaji (Gujrat), India. *(Rajkumar Ghosh)*

FIGURE 2.11 **Sigmoidal shaped ductile sheared quartz vein within meta-greywacke showing sinistral shear sense associated with en-echelon shear fractures.** Kumbhalgarh Formation, ~85 km northwest of Udaipur, Rajasthan, India. *(Swagato Dasgupta)*

FIGURE 2.12 **This is type 1 flanking structure within calc schist.** Quartz vein crosscutting element top-to-left sheared. At bottom right, the tail of the sheared quartzite body boudinaged. Taleti Village: (Latitude: 24° 21.792′N Longitude: 72° 54.640′E), Ambaji, Gujrat, India. See Passchier (2001), Mukherjee (2007, 2011a, 2013a, 2014a,b), Mukherjee and Koyi (2009) and Koyi et al. (2013) for detail of flanking structures. *(Rajkumar Ghosh)*

FIGURE 2.13 **Around m-scale delta-type asymmetric-feldspar clast.** See Mukherjee (2010a,b) for general term "fish" to describe such structures. Asymmetric tails containing quartz of much reduced grain size indicate top-to-left shear. Chitrial granite; Place: Chitrial Village, Nalgonda, India. Delta structures are usually found under microscales from ductile shear zones. Very rarely, they are found in mesoscale as well (e.g., Figure 4 in Mukherjee (2013b). *(Rajkumar Ghosh)*

FIGURE 2.14 **Top-to-right/west sheared intercalations of quartzites and phyllites showing antiformal thrust stacks at the Udayagiri Durgam Hill, Nellore Schist Belt.** Kondamadipalle village (14°53′37.54″N, 79°16′23.20″E), Nellore District, Andhra Pradesh, India. *(Sankha Das)*

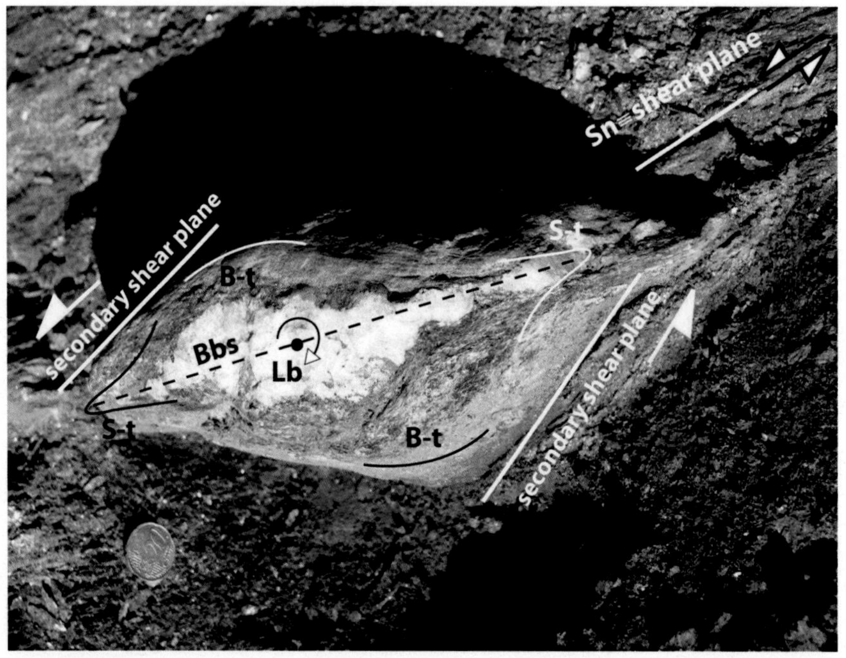

FIGURE 2.15 **A classic shearband boudin generated by deformation of a thicker tabular competent body (quartz vein) within in a less competent matrix.** This quartz vein does not exhibit internal folding. The methodology of analysis of shearband boudins proposed by Pamplona and Rodrigues (2011) includes geometrical parameters that allow a better characterization of this kind of geological bodies: Bbs-orientation of the boudin symmetry plane, Lb-boudin axis: defined by the intersection points between the opposite sharp tips (S-t) and the secondary synthetic shear plane; the blunt tip (B-t) that is the zone of boudin surface that corresponds to the development of a convexity at the interface boudin/matrix with secondary synthetic shear plane. Criteria to kinematic interpretation as a bulk sinistral shear zone using shearband boudin parameters: Bbs rotates dextral (antithetic) relatively to foliation plane (Sn); when looking parallel to Bbs, at a sharp tip of the boudin, toward the interior of the boudin, the B-t is on the left-hand. Photo scale is given by the coin (diameter = 22.25 mm). Outcrops of Salgosa Sector (NW of Portugal, 4642140N 524687E, WGS84). See Pamplona and Rodrigues (2011) and Pamplona et al. (2014) for details. Shearbands/secondary shears are considered readily by structural geologists as evidence of general shear deformation (Mukherjee and Koyi, 2010b; Mukherjee, 2013c). *(Jorge Pamplona, Benedito Calejo Rodrigues, Carlos Fernández)*

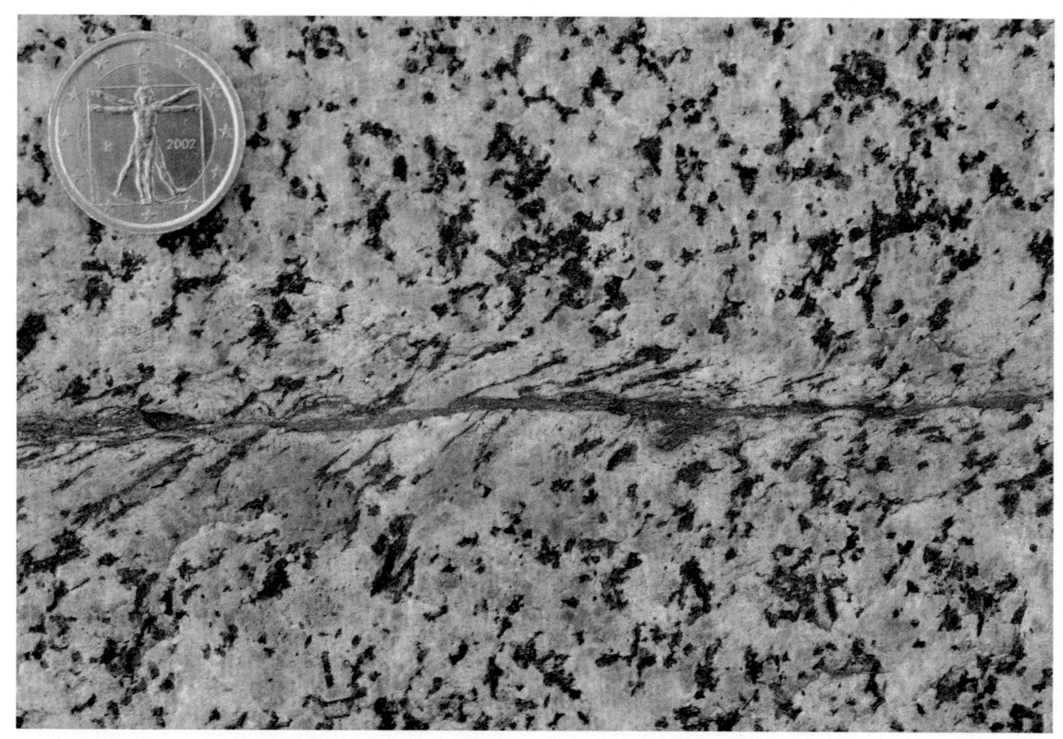

FIGURE 2.16 **Incipient heterogeneous ductile shear zone localized on a precursor fracture.** The discrete fracture is still clearly visible in the shear zone core. The new foliation in the shear zone is well defined by the elongation of deformed pseudomorphs after original magmatic biotite. Precursor fractures can run for tens of meters and typically have a consistent stepping geometry (left stepping for dextral overprinting shear zones and vice versa) suggesting that the fractures developed in a kinematic framework similar to the subsequent ductile shear. Metamorphic minerals biotite and garnet concentrated in the fracture are also consistent with Alpine upper amphibolite facies conditions during ductile shear. The sense of shear is dextral. This and the three following photographs are from the Neves area of the Tauern window in the Eastern European Alps. Superb exposure is provided by the recent retreat of the Mesule glacier, exposing a perfectly polished and subhorizontal pavement at the foot of the glacier, at ~2600-m height above sea level, The investigated shear zones are subvertical, with a flat-lying mineral stretching lineation, so that the horizontal surface provides a profile section. Shear zones are developed in a pre-Alpine pluton of dominantly granodioritic composition, which was metamorphosed during Alpine time under upper amphibolite facies conditions. The shear zones developed at near-peak Alpine metamorphic conditions. Details of the location, regional geology and a more extensive description of the structures are available in Mancktelow and Pennacchioni (2005), Pennacchioni and Mancktelow (2007) and Mancktelow and Pennacchioni (2013). Coin (1 euro) for scale. *(Neil Mancktelow, Giorgio Pennacchioni)*

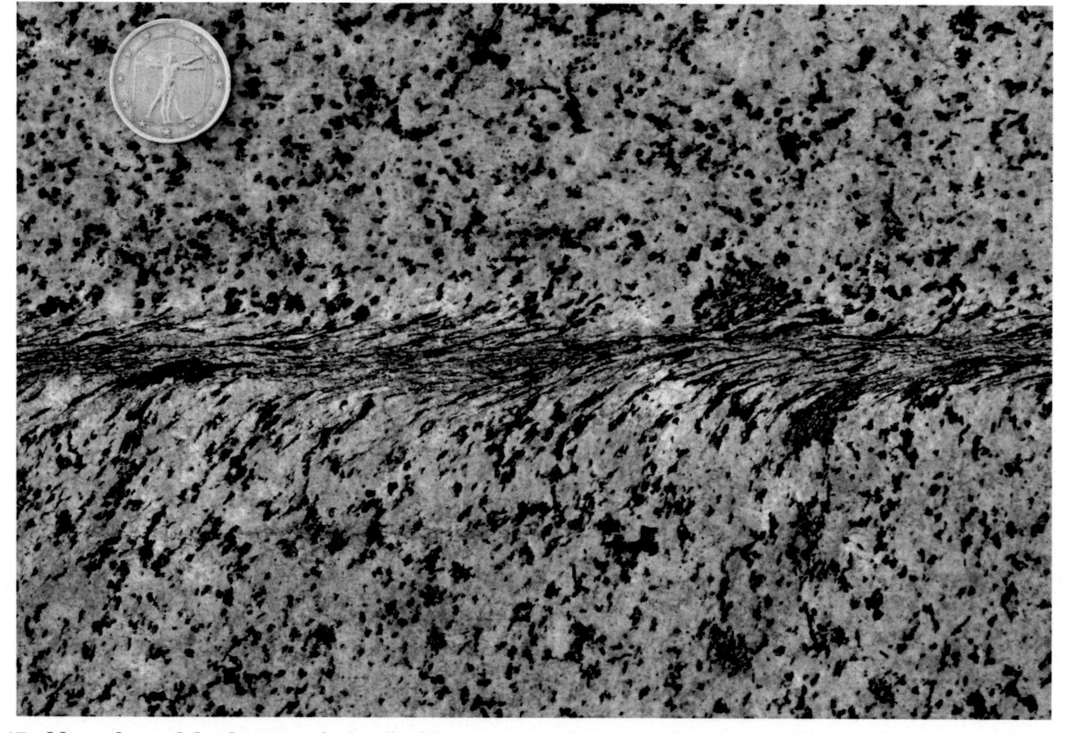

FIGURE 2.17 **More advanced development of a localized heterogeneous shear zone.** Here the initial fracture is no longer visible but the outcrop pattern of these ductile shear zones on the meter to tens of meter scale still mimics the original fracture pattern. Dextral shear. Coin (1 euro) for scale. *(Neil Mancktelow, Giorgio Pennacchioni)*

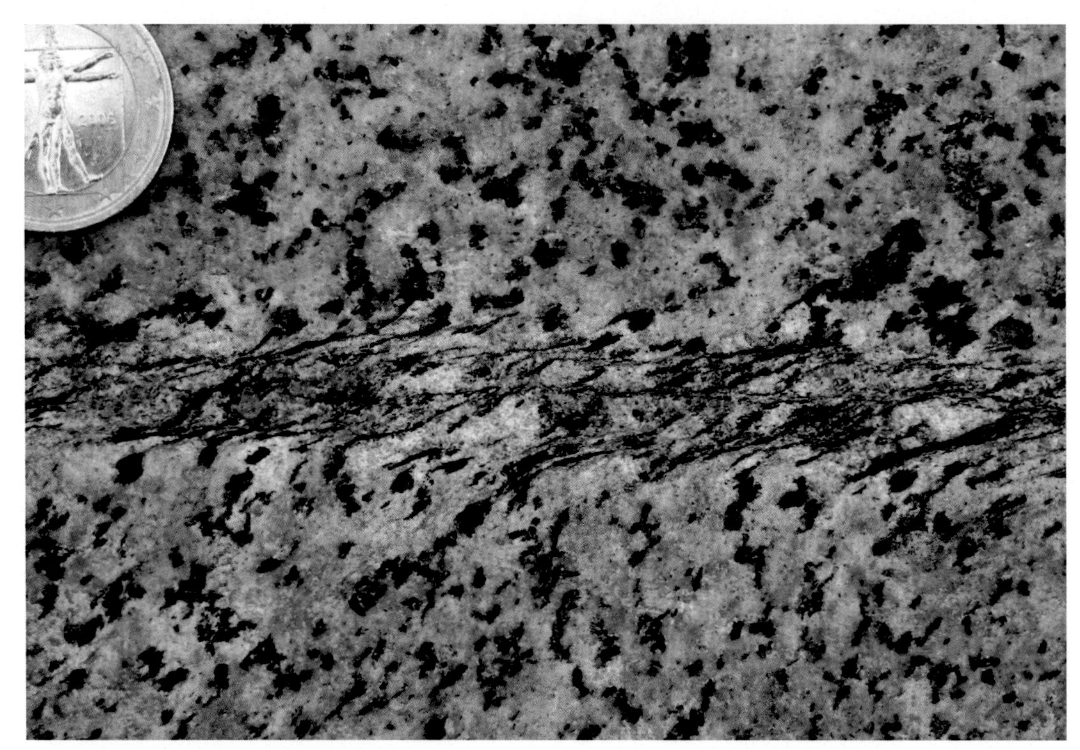

FIGURE 2.18 **A pair of incipient localized dextral shear zones developed to either side of a precursor facture.** Slip occurs both on the central fracture and on the flanking pair of heterogeneous shear zones. Coin (1 euro) for scale. *(Neil Mancktelow, Giorgio Pennacchioni)*

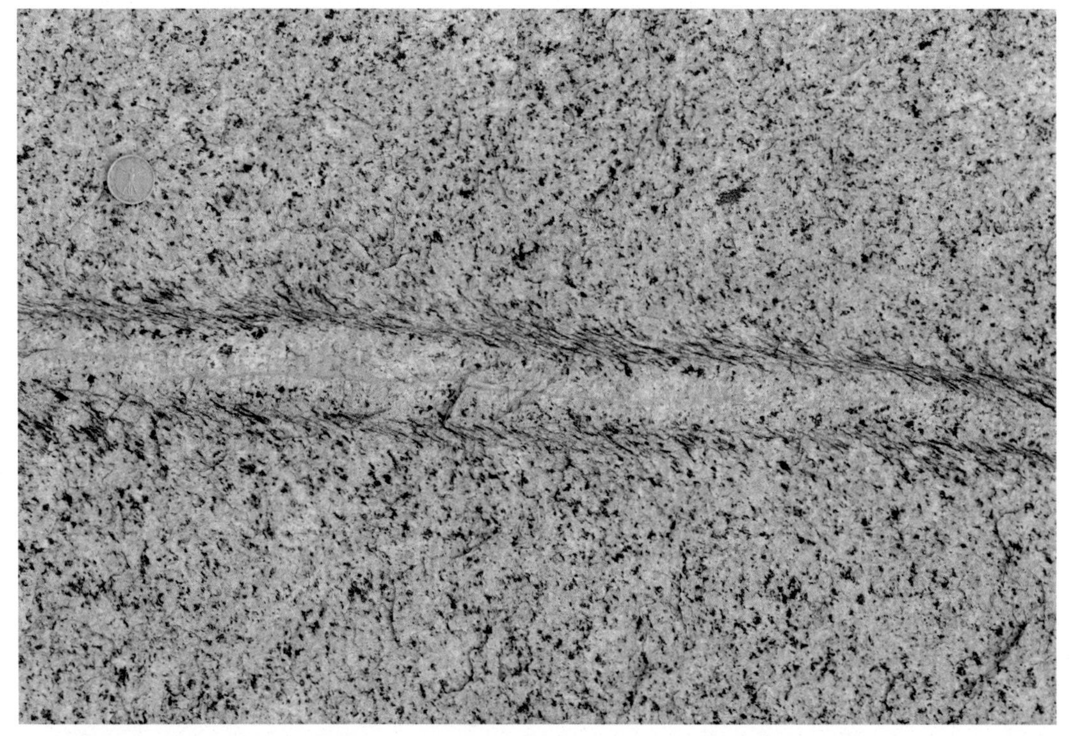

FIGURE 2.19 **Typical paired shear zones developed approximately symmetrically to either side of an initial fracture.** The fracture itself has filled to form an epidote-garnet vein and fluid-rock interaction to either side of the fracture has produced a bleached zone enriched in plagioclase and epidote and depleted in biotite. The paired shear zones flank this bleached zone and must have developed after the altered zone reached its current width, rather than during its progressive development and broadening. Similar paired shear zones develop at the boundary of relatively strong layers such as aplite dykes within the granodiorite. This suggests that paired shear zones develop by reactivating boundaries of stronger layers (e.g. the bleached halo here). Sinistral shear. Coin (1 euro) for scale. *(Neil Mancktelow, Giorgio Pennacchioni)*

FIGURE 2.20 **A polished slab of granite mylonite from Shirakami Mountains of N Japan (40°27′27.7′N, 139°57′7.8′E).** This slab was cut parallel to the lineation and perpendicular to the foliation. Porphyroclasts of K-feldspar develop with pressure shadows. Foliation made by elongated aggregates of fine-grained biotite (dark parts) and quartz ribbons (pale gray lenticular parts) is clear on the slab. Shear bands (C-planes) develop oblique to the main foliation (S-plane): sinistral ductile shear. Scale bar: 10cm. The granite mylonite originated from biotite granite of the Shirakami-dake Granites. Its K-Ar hornblende, biotite ages are ~90Ma (Fujimoto and Yamamoto, 2010). Mylonitization occurred ~90Ma under upper greenschist to amphibolite facies condition. This mylonite zone in the Shirakami Mountain is a possible northern extension of the shear zone—Tanagura Tectonic Line, which is a possible NE extension of the Median Tectonic Line (Takahashi, 2002). Therefore, the mylonite zone is one of the key areas to study the Cretaceous tectonics of E Asia. *(Yutaka Takahashi)*

FIGURE 2.21 **Boudinaged quartz veins within banded magnetite quartzite and magnetite garnet-biotite schist intercalation.** Top-to-right ductile shear and synthetic secondary shear present. Tiranga hill, Pur area, Bhilwara, Rajasthan, India. See Mukherjee and Koyi (2010b) for microboudins. *(Ranjan Gupta)*

FIGURE 2.22 **Photomicrograph shows a duplex structure in a strike slip fault system.** Shear planes in the microduplex are defined by fine amphibole grains and are marked by the broken white line. The white arrows indicate the sense of shear movement. The rock is granodiorite from Kajalbas area (hanging wall side) nearly a kilometer east of Phulad Shear Zone (Ghosh et al., 1999; Sengupta and Ghosh, 2004) of the Southern Delhi Fold Belt, Rajasthan, India. The rock shows a prominent foliation parallel to the mylonitic foliation of the shear zone and is dissected by late brittle-ductile shear bands. Photo in cross-polarized light. The width of the photo is 40 mm. Sample location is 25°38′58.2″N, 73°49′48.9″E. *(Sadhana M. Chatterjee, Sudipta Sengupta)*

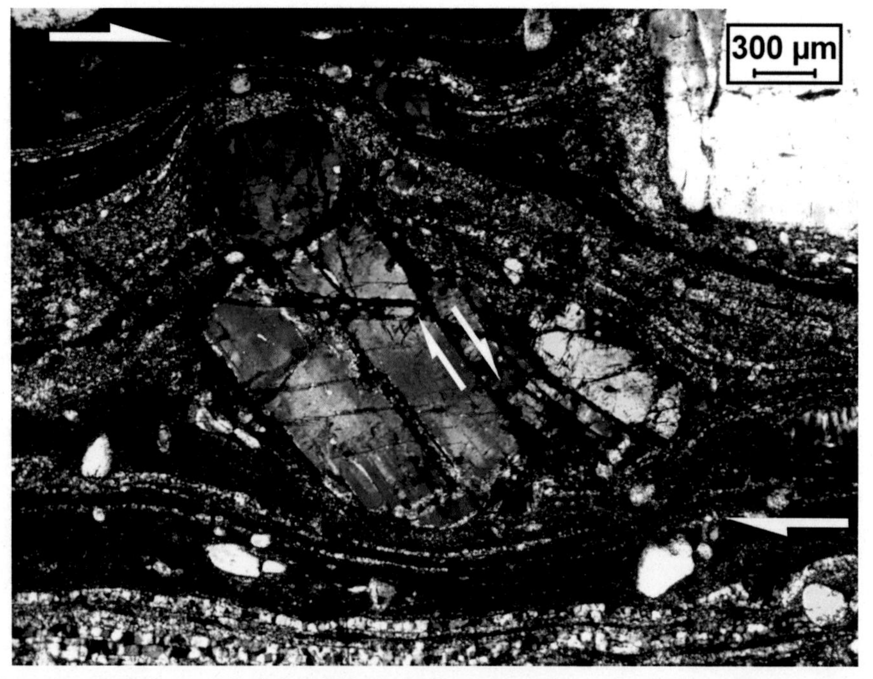

FIGURE 2.23 **The photomicrograph shows a plagioclase feldspar porphyroclast in fine-grained recrystallized matrix of quartz, feldspar, and calcite.** The sample is a quartzofeldspathic mylonite. The feldspar porhyroclast in the center is divided into four pieces by grain scale faults and shows a typical bookshelf sliding (Ghosh, 1993, p. 251) or fragmented domino type feature (Trouw et al., 2010). The fractures in the feldspar megacryst are at an angle with the general mylonitic foliation. The sense of slip along the fractures (shown by small white arrows) is sympathetic to the dextral shear direction of the mylonite (shown by larger arrows). However, the grain shows antithetic rotation indicating transpressional deformation (compare with Figure 21.13(b) of Ghosh (1993), p. 521). The megacryst in the upper right side shows horizontal and vertical fractures. There is no slip along these fractures. Photo in cross-polarized light. Sample location: 25°36′33.1″N, 73°48′45.1″E, from Phulad Shear Zone, Rajasthan, India (Sengupta and Ghosh, 2004). *(Sudipta Sengupta, Sadhana M. Chatterjee)*

FIGURE 2.24 **Post D_2, syn-D_4 ultramylonite shear zone in augen gneiss, Bastar Craton, Sambalpur, India.** The gneissic foliation (S_2) was dextrally sheared and transposed by a late ultramylonitic, amphibolite facies shear zone (dark gray). The ultramylonitic shear foliation, with 81/83°N latitude, is synchronous with the Eastern Ghats Boundary Fault (Bhadra et al., 2004; Biswal et al., 2000, 2007; Dobmeier and Raith, 2003). S of Sambalpur town, Odisha, India. *(Arindam Dutta, Saibal Gupta, M.K. Panigrahi)*

FIGURE 2.25 **Asymmetric quartz porphyroclasts in sheared quartzofeldspathic gneiss indicate an apparent dextral shear.** The steeply SW dipping mylonitic foliation parallels the WNW-ESE trending Kerajang Fault that separates the Rengali Province from the Eastern Ghats Mobile Belt (Crowe et al., 2003; Dutta et al., 2010) along the National Highway-6, E of Jamankira, Odisha, India. *(Arindam Dutta, Saibal Gupta, M.K. Panigrahi)*

FIGURE 2.26 **Top-to-SW sheared sigmoid quartz veins.** Event of extension followed by ductile shear. Centimeter scale product of Himalayan compression. Lesser Himalaya, 31°14.578′N, 76°58.96′E, Mangu village, district: Solan, Himachal Pradesh, India. *(Tuhin Biswas)*

FIGURE 2.27 **Photograph of a dilational jog within the Chrystalls Beach Complex, Otago, New Zealand.** The Chrystalls Beach Complex is an accretionary mélange, comprising lenses of sandstone, chert, and minor basalt within a cleaved mudstone matrix containing an anastomosing vein network. This sheared rock assemblage forms part of the Otago schist, and was metamorphosed at sub-greenschist facies during Triassic–Jurassic subduction under the Gondwana margin (Hada et al., 2001). The photograph is taken looking W at a subvertical face. Thus the dilational jog indicates top-to-north shear (top-to-right in figure). This is also typical of the mélange as a whole. The tensile veins within the jog and the enveloping slickenfiber-coated shear surfaces are nearly orthogonal, intersecting at an angle of ~80°. The presence of tensile veins implies that fluid pressure exceeded the least compressive stress, at least locally within the jog. In addition, the high angle between shear and extension fractures requires high fluid pressure and low frictional resistance along the shear surfaces for their continued reactivation (Fagereng et al., 2010). Internally, the slickenfiber shear veins (as seen on the left in the photograph) comprise shear surfaces at a low angle to the vein margin, separated by incrementally grown quart layers. These shear surfaces were therefore active episodically, possibly governed by fluid pressure cycling, leading to episodic growth of dilational jogs (Fagereng et al., 2011). This "dilational hydro-shear" may be analogous to repeating, low stress-drop, microearthquakes observed in actively deforming accretionary complexes. *(Ake Fagereng)*

FIGURE 2.28 **A polished slab of the augen gneiss/granite mylonite in Funatsu Shear Zone within the Hida Belt of Southwest Japan (36°38′34.8′N, 137°33′40.3′E) (Harayama et al., 2000).** This slab was sectioned parallel to lineation and perpendicular to foliation. Porphyrocrasts of K-feldspar are characteristic with anisotropic pressure shadows: dextral sheared. Foliation made by elongated aggregates of fine-grained biotite: dark parts, and "quartz ribbons": pale gray lenticular parts, is clear on the nearly horizontal surface. Scale bar: 10 cm. The Funatsu Shear Zone is a late Triassic dextral ductile shear zone in the Hida Belt, which is a possible NE extension of the Cheongsan Shear Zone in the Korean Peninsula (Takahashi et al., 2010). *(Yutaka Takahashi)*

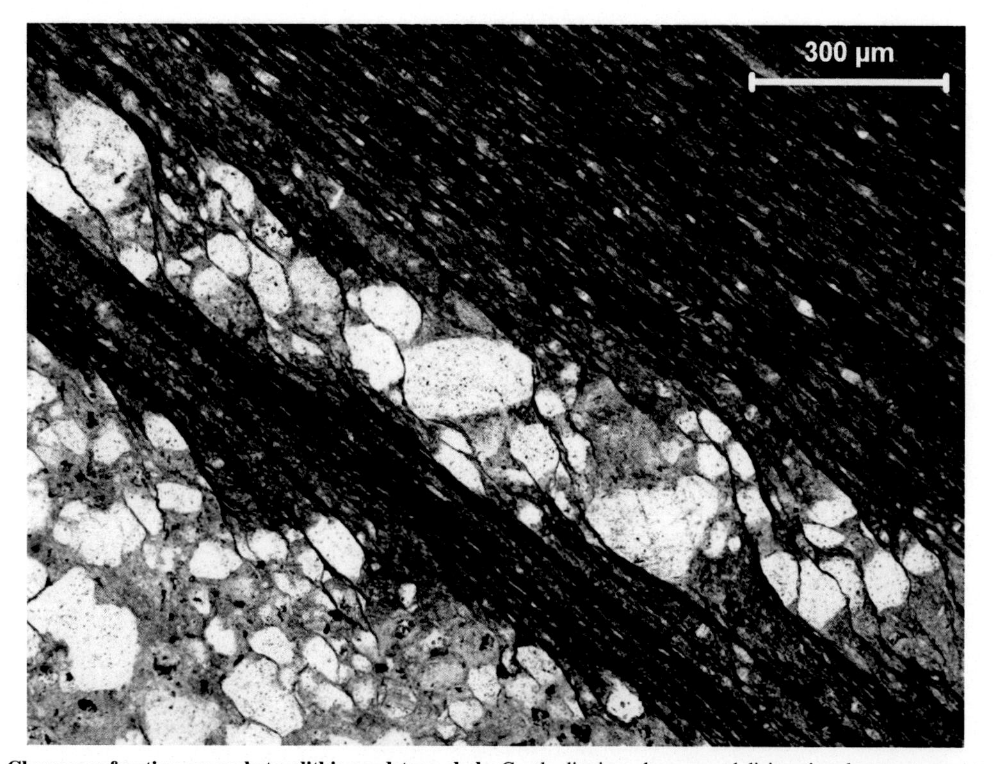

FIGURE 2.29 **Cleavage refraction across heterolithic sandstone-shale.** Gently dipping, close-spaced disjunctive cleavage to penetrative slaty cleavage (Engelder and Marshak, 1985) in pelitic laminae refracts to steeper spaced cleavage in the psammitic laminae. The cleavage seams, darker bands, are associated with pressure solution (Rutter, 1983) as evidenced by sharp truncation of detrital quartz grain margins. Note that cleavage spacing in psammitic laminae is controlled by the size of quartz clasts, which show little intracrystalline deformation. Photo in plane polarized light. Kurnool Group rocks. Palnad, Southern India. Bar scale: 300 μm. *(Dilip Saha)*

FIGURE 2.30 **Sigmoid brittle P-planes bound by Y-planes in mafic schist: top-to-right shear.** Foliation boudins below. 4-mm pen for scale. Rampur *meta*-basalt, Inner Lesser Himalaya, Sutlej Valley, Himachal Pradesh, India. *(Subhadip Mandal)*

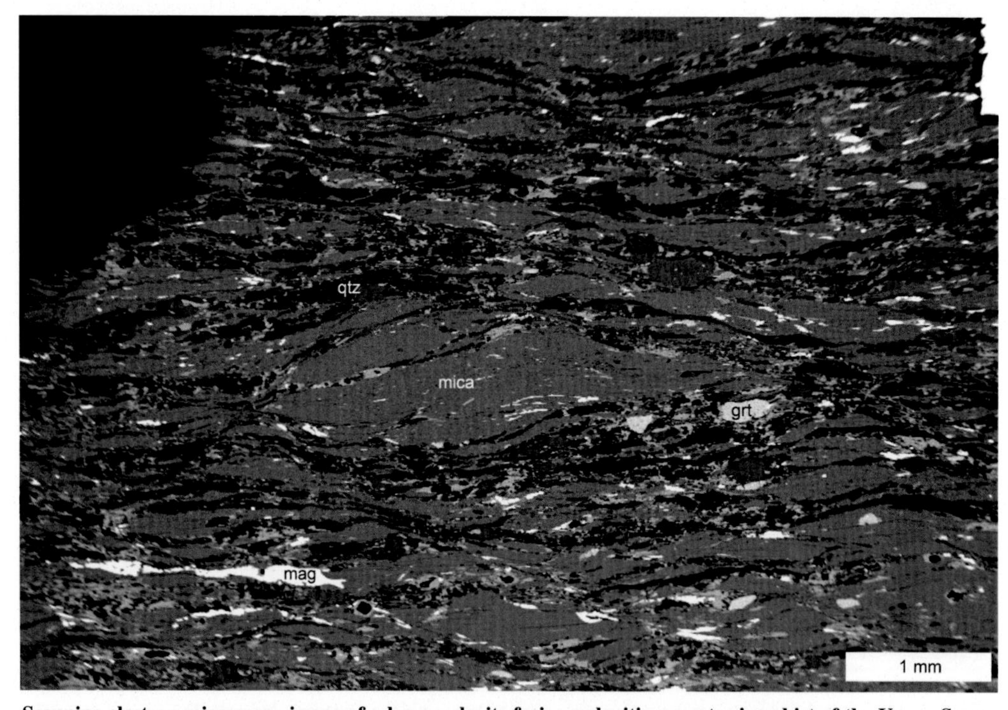

FIGURE 2.31 **Scanning electron microscope image of a lower eclogite facies mylonitic garnet-micaschist of the Upper Seve nappe with kinematically grown magnetite (mag) in a quartz (qtz)-mica-feldspar-matrix with mica fish indicating top-to-right sense of shear (i.e. hinterland-directed) on top of an extrusion wedge in the central Scandinavian Caledonides** (Grimmer et al., 2015; Kontny et al., 2012). Note also developed synthetic shear bands. *(Jens Carsten Grimmer)*

FIGURE 2.32 **Mylonitized micaschist shows a top-to-right sheared biotite "fish."** Cleavage planes are at an angle to ductile shear. Therefore it is an r-parallelogram fish of Mukherjee (2011b). As per a different classification scheme, this is a Group 3 fish (ten Grotenhuis et al., 2003). See Mukherjee (2012) for simple shear mechanism. Natyal (23°05′39.51″N-73°42′06.63″E), Kadana Formation, Lunawada Group, Gujarat, India. *(Aditya Joshi, M.A. Limaye, Bhushan S. Deota)*

FIGURE 2.33 **Asymmetric sheared granite pebble in mylonitic garnet-micaschists of the Middle Allochthon Seve Nappe Complex indicates foreland-directed (top-to-left) tectonic transport on top of the (not exposed) Ammarnäs Complex** (Grimmer et al., 2011). *(Jens Carsten Grimmer)*

FIGURE 2.34 **Sigma clasts, S-C structures in soft-sediment.** Microstructures in ~1-m thick clastic infill of a steeply dipping synsedimentary fault displaced Mesozoic rocks. Compositional bands in clastic infill parallel the fault. The major structures: calcite shear veins correspond to stacked slickenside fibers which occur along minor faults within the infill and parallel to main fault. Calcite fibers alter with millimeter-thick bands of clastic sediment (quartz and chert grains, clays with iron impregnation). Sigma clasts with calcitic tails indicate top-to-left shear. The quartz grains within the sigma clasts are intensely fractured and extended by calcite veinlets, which are also consistent with sinistral shear. In the lower part, parallel iron-impregnated dark clays bands correspond to S-foliation planes and merge with C-shear surfaces parallel to calcitic shear veins. Thus ductile S-C structure (Platt and Vissers, 1980) is defined. All these features are geometrically similar to structures found in crystal-plastically deformed metamorphic rocks. However, the sediment infill and host rocks are unmetamorphosed and were never buried below 2 km. The interpretation is that the structures formed in unconsolidated state of the clastic infill, during the burial path of the sediment. The whole deformed zone is crosscut by the youngest parallel set of calcite veinlets, which could form after the complete diagenesis of the shear zone rocks. Middle Eocene sandstone, siltstone. Location: Vöröshíd quarry, Tardos village, Gerecse Hills, Hungary. Coordinates: 47°41′48.47″N, 18°26′56.25″E. *(László Fodor)*

FIGURE 2.35 **A top-to-left ductile sheared sigmoid quartz lens in kyanite-garnet gneiss, hanging wall rock of the Main Central thrust.** Location: Unit I, Greater Himalaya, Kali-Gandaki section, Central Nepal Himalaya. *(Subodha Khanal)*

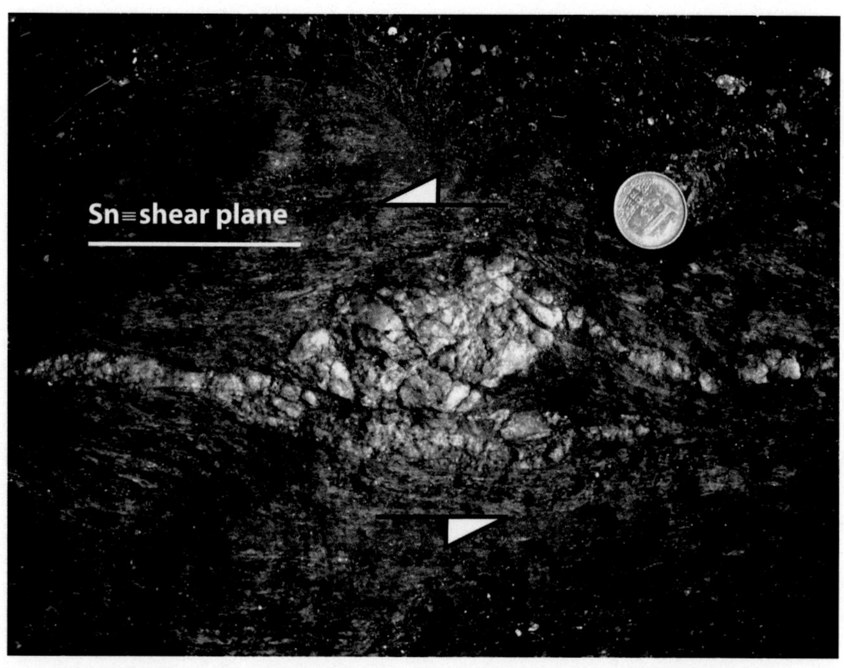

FIGURE 2.36 **The continuous flattening of a folded package, rather than the developing of new shear structures, develops a false sigma feature called fold-boudin.** Folding starts with buckling of quartz veins with the axial plane oblique to shear plane. Then the axial plane rotates and the fold tightens. At this stage, the position of hinges still records an asymmetric distribution relative to the shear plane. This fold train exhibits an apparently antithetic kinematics relative to the shear plane given by its external morphology. The key structure to an appropriate kinematic interpretation is the internal symmetry of the folds. After the folds tighten, the fold train starts to behave as one single unit that rotates due to the flow vorticity. Diameter of the coin as marker is 23.25 mm. Outcrops of Salgosa Sector (NW of Portugal, 4642140N 524687E, WGS84). See Pamplona and Rodrigues (2011) and Pamplona et al. (2014) for details. *(Jorge Pamplona, Benedito Calejo Rodrigues, Carlos Fernández)*

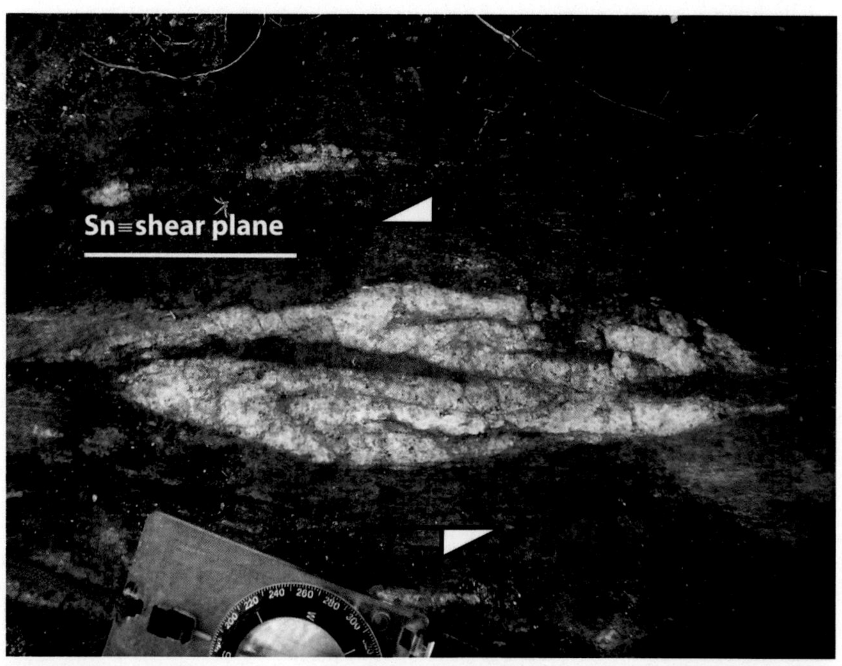

FIGURE 2.37 **The evolutive sequence of fold-boudin generates a thick compact, parallel-sided body designed by stacked-fold-boudin.** The hinges and axial planes of folds tend to parallel the shear plane. At this and the previous stage, it is possible to find remains of old ductile matrix (amphibolitic facies of quartz micaschist) interbedded with quartz fold flanks. The condition to achieve this evolutionary level is verified when the inhibition of shear rupture predominates, which develops a localized central folding on the vein, generating a fold-boudin with two narrow and opposite tails. Photo scale is given by compass. Outcrops of Salgosa Sector (NW of Portugal, 4642140N 524687E, WGS84). See Pamplona and Rodrigues (2011) and Pamplona et al. (2014) for details. *(Jorge Pamplona, Benedito Calejo Rodrigues, Carlos Fernández)*

FIGURE 2.38 **When stacked-folds-boudins are fully developed, the geologic body—originally a thinner vein—acquired the critical thickness that develop the shearband boudin.** The internal structure is composed by successive package of folds flanks while external morphology shows typical shear band structures like a sigmoid with the tips bound by opposite secondary shear planes. Photo scale is given by compass. Outcrops of Salgosa Sector (NW of Portugal, 4642140N 524687E, WGS84). See Pamplona and Rodrigues (2011) and Pamplona et al. (2014) for details. *(Jorge Pamplona, Benedito Calejo Rodrigues, Carlos Fernández)*

FIGURE 2.39 **Top-to-S/SW ductile sheared parallelogram/sigmoid mud unit inside sandstone.** Competency contrast between sandstone and mud controlled the shape of the mud unit (like Treagus and Lan, 2003). Mohand, Siwalik Himalaya, Roorkee-Dehradun transect, Uttarakhand, India. *(Tuhin Biswas, Soumyajit Mukherjee)*

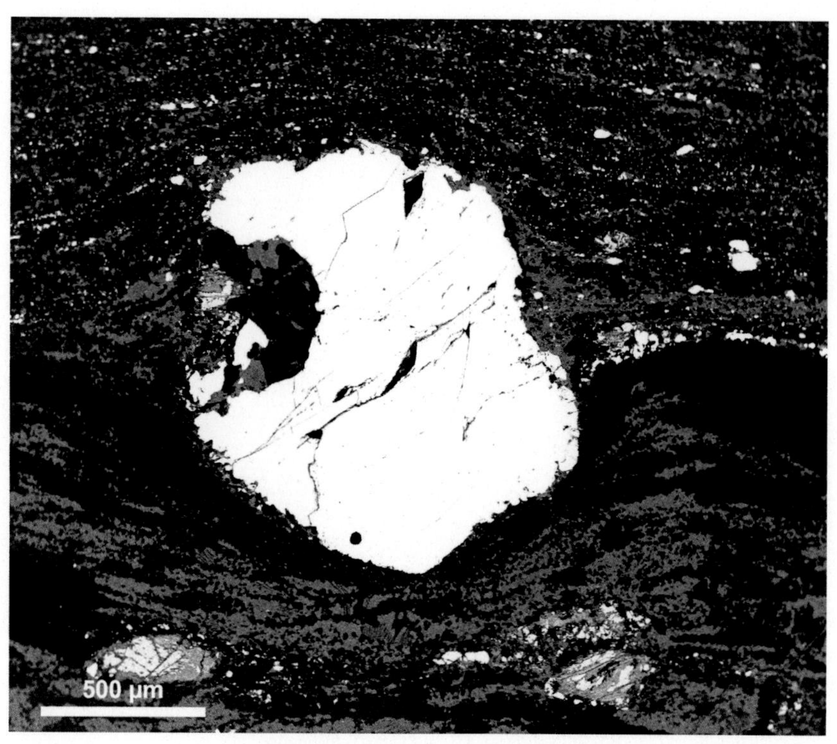

FIGURE 2.40 **Backscattered electron image of a titanite porphyroclast in an ultramylonitic shear zone.** The shear zone is developed in a coarse-grained metasyenite from the boundary between the Adirondack Highlands and the Adirondack Lowlands (New York, USA). The rounded porphyroclast is consistent with rotation of the porphyroclast relative to the grain-size reduced matrix. The embayment within the porphyroclast contains coarser matrix grains that were apparently protected from further grain-size reduction within the strain shadow of the surrounding titanite. Foliation is defined by very fine-scale compositional layering that formed by extreme attenuation of individual dynamically recrystallized quartz, alkali feldspar, and augite grains. The matrix in the upper half of the image has undergone greater attenuation and/or a greater degree of phase mixing during shearing than the matrix in lower half of the image. Coarse feldspar grains in the nearby unsheared metasyenite are perthitic, but dynamic recrystallization separated the exsolved domains into individual sodic or potassic matrix grains in the shear zone. Darkest gray in the image is quartz; intermediate grays are potassic and sodic feldspars. The image plane is perpendicular to foliation; no lineation is observed in outcrop or hand sample. *(Chloë Bonamici)*

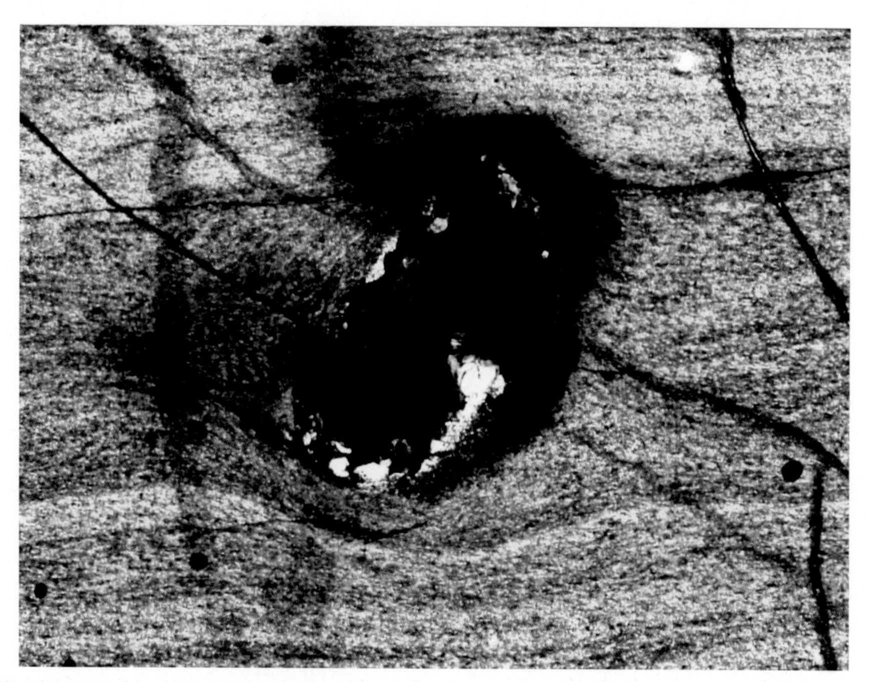

FIGURE 2.41 **Delta-like structure in an ultramylonitic peridotite.** The partially recrystallized porphyroclast is composed of spinel. The wings consist of very fine spinel grains and they are very thin. Plane Polarized Light. Location: Archipelago of Saint Peter and Saint Paul (Brazil). Width of view: 2.5 mm. *(Suellen Olívia Cândida Pinto, Leonardo Evangelista Lagoeiro, Luiz Sérgio Amarante Simões, Paola Ferreira Barbosa)*

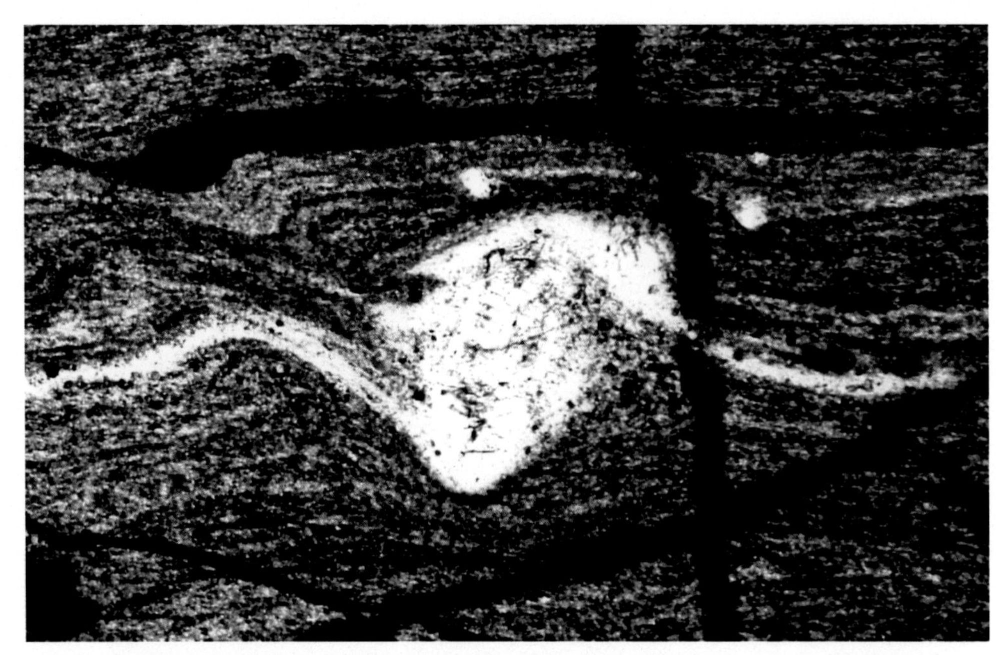

FIGURE 2.42 **Delta (δ) microstructure of olivine.** A porphyroclast of forsterite within peridotitic has asymmetric wings of fine recrystallized grains. The core is recrystallized almost entirely. The wings are of delta type: top-to-left sheared. The asymmetric fold in the NW quadrant too provides a consistent shear sense. Plane polarized light. Location: Archipelago of Saint Peter and Saint Paul (Brazil). Width of view: 0.25 mm. *(Suellen Olívia Cândida Pinto, Leonardo Evangelista Lagoeiro, Luiz Sérgio Amarante Simões, Paola Ferreira Barbosa)*

FIGURE 2.43 **Top-to-right ductile sheared syntectonic microscopic porphyroblast of chloritoid (cld) under cross-polars.** Asymmetric pressure shadow of quartz around the porphyroblast bound by mica grains. Middle greenschist facies metamorphosed Proterozoic chloritoid bearing garnet-muscovite schist, Udayagiri Group, Nellore schist belt. Arlapadia village (15°02′08″N, 79°16′02″), Prakasam District, Andhra Pradesh, India. *(Sankha Das)*

FIGURE 2.44 **Top-to-right ductile sheared microscopic garnet with spiral, early foliations (S-internal: Si) within the core.** See Mukherjee (2014c) for review of inclusion minerals. The rim is almost free of inclusions. This indicates that the grain grew in two stages. After its initial syntectonic growth, the garnet grain grew again in a post-tectonic phase. Middle greenschist facies metamorphosed Proterozoic chloritoid bearing garnet-muscovite schist, Udayagiri Group, Nellore schist belt. Arlapadia village (15°02′08″N, 79°16′02″), Prakasam District, Andhra Pradesh, India. *(Sankha Das)*

FIGURE 2.45 **Top-to-right ductile sheared syntectonic microscopic garnet under plane polarized light.** Middle greenschist facies metamorphosed Proterozoic chloritoid bearing garnet-muscovite schist, Udayagiri Group, Nellore schist belt. West of Peddarajupalem Village (15°06′38.00″N, 79°15′44″E) Prakasam District, Andhra Pradesh. *(Sankha Das)*

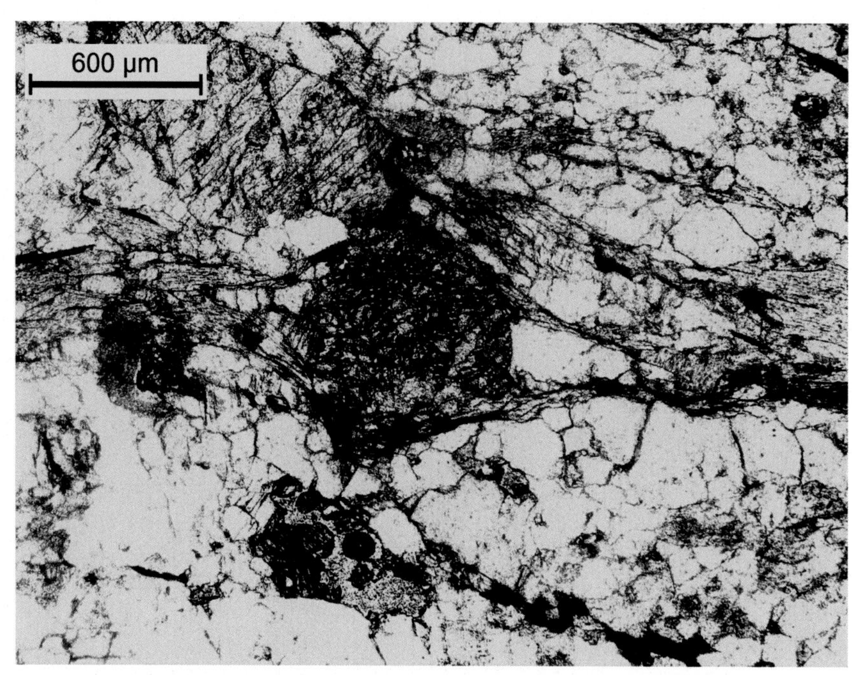

FIGURE 2.46 **Fractured garnet porphyroblast showing phi-structure and inter-tectonic deformation.** No shear sense indicated. Plane polarized light. Photo width: 1.5 mm. Tso Morari Crystallines, Ladakh, India. *(Kankajit Maji)*

FIGURE 2.47 **Photomicrograph of biotite + kyanite-bearing micaschists from the Num orthogneiss in the Makalu valley (East Nepal).** Plane polarized light, width of view *c.* 4 mm. The main foliation, oriented parallel to the long side of the picture, is highlighted mainly by a strong shape-preferred orientation of polycrystalline quartz ribbons, kyanite crystals (with high relief and light gray-yellowish color) and less than 1-mm thick-biotite layers. Quartz grains have (<500 µm) and show a weak shape-preferred orientation (generally parallel or at very low angle to the main foliation), weak undulatory extinction and straight to interlobate grain boundaries suggesting high-temperature dynamic recrystallization mechanism (Grain Boundary Migration) probably associated with annealing. Biotite crystals have no evidences of retrograde replacement. Kyanite crystals show inequigranular grain size (ranging from 1 mm to <100 µm) and {100} and {001} sections oriented parallel and orthogonal to the main foliation respectively. In particular, in the center of the picture, a large kyanite crystal showing euhedral basal section is wrapped by the main foliation made of biotite and strongly elongated kyanite crystal that shows aspect ratio of *c.* 20:1 and a well-developed cleavage system. Although geometric relationships could indicate two different generations of kyanite (pre- and syn-foliation), the bending of the elongated crystal may be also explained considering a differential growth velocity between {100} and {001} (basal) crystal sections during the same synkynematic metamorphic blastesis. *(Chiara Frassi)*

FIGURE 2.48 **Photomicrograph of asymmetric quartz ribbon from the mylonitic micaschists in the Variscan basement in Northern Sardinia Island (Italy).** The picture is viewed with the lambda plate under crossed polarizer light. Width of view *c.* 7 mm. The main mylonitic foliation is defined by quartz, fine-grained garnet, feldspar, and very thin-layers of muscovite and biotite aggregates (the irregular layers with dark red-violet color). A < 1 cm-thick shear zone (upper left to lower right) deforms the main foliation and a mm-thick quartz ribbon suggesting a dextral sense of shear. In the left and right portions of the ribbon, quartz shows interlobate grain boundaries and seriate grain size indicating Grain Boundary Migration recrystallization mechanism. Within the shear zone in the thinner portion of the ribbon, quartz grains are small, equigranular and slightly elongated indicating that intracrystalline deformation probably occurred by Subgrain Rotation recrystallization. The strong crystal preferred orientation and the well-defined grain shape foliation suggest a sense of shear opposite to those inferred by other microstructures (included the asymmetry of quartz ribbon itself). This opposite kinematics can be interpreted as the result of rigid body rotation on quartz grains affected by low-velocity intracrystalline deformation. *(Chiara Frassi)*

FIGURE 2.49 **S-C-fabrics in antigorite (atg) mylonite and asymmetric sheared composite olivine-pyroxene (ol, opx) clasts indicate top-to-right tectonic transport.** Cross-polars. Seve Nappe Complex, Central Scandinavian Caledonides, Sweden. *(Jens Carsten Grimmer)*

FIGURE 2.50 **Photomicrograph of mylonitic peridotite cropping out in the northwestern Elba island (Italy).** Plane polarized light, width of view *c.* 4 mm. The main foliation, oriented parallel to the long side of the picture, is marked by a strong shape-preferred orientation of olivine (the high-relief colorless minerals with microfractures orthogonal to the main foliation), spinel (the dark brown elongated minerals) and amphibole with minor amount of phlogopite, pyroxene, and plagioclase. The foliation is the result of a progressive shearing developed during Mesozoic mantle exhumation. The earlier metamorphic paragenesis is defined by olivine (Fo_{70-80}), magnesium–aluminum spinel, rare anorthite (the bright colorless elongated crystal in the central lower portion of the picture), phlogopite and very small orthopyroxene crystals (10–30 μm). Olivine crystals generally show elongated shape (see the olivine ribbon in the lower right portion of the picture). However, the olivine porphyroclast in the center-left portion of the picture shows a very small aspect ratio and sigma-type geometry with asymmetric tails indicating a top-to-the right sense of shear. The tails are made of strongly elongated olivine and amphibole crystals grew during the second metamorphic event. The amphibole (Mg-hornblende/tremolite) is highlighted by yellowish to pale brown acicular to fibrous crystals. Finally, small amphiboles with gedrite-anthophyllite composition (see the small colorless rhombohedric crystals crosscutting the main foliation) grew during the Late Miocene contact metamorphism related to emplacement of the Mt. Capanne intrusion. *(Chiara Frassi, Giovanni Musumeci, Francesco Mazzarini)*

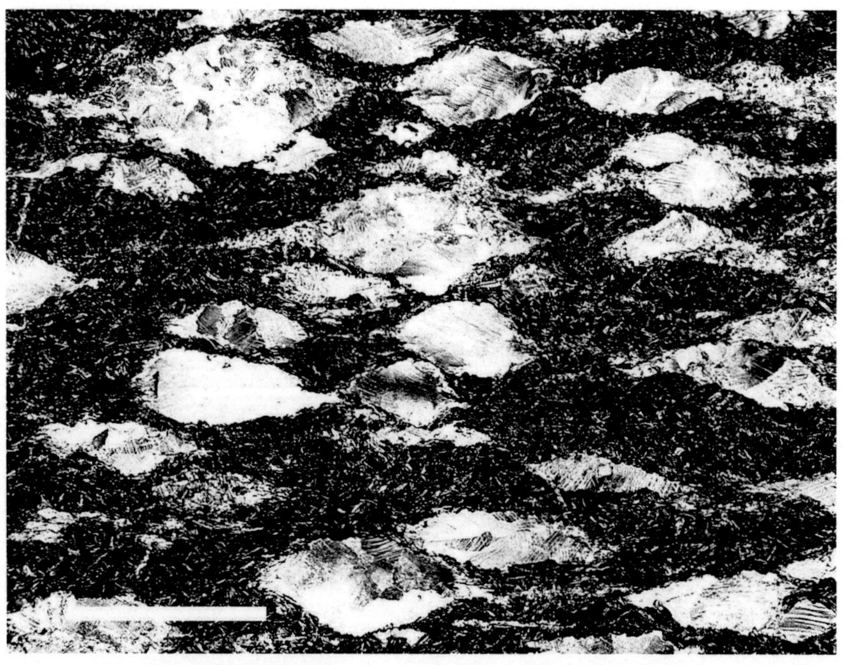

FIGURE 2.51 **A thin section of a basic lava containing deformed amygdules.** The sample was taken from the Manba Unit at a Jurassic accretionary complex in the Northern Chichibu Belt, Central Japan (Shimizu and Yoshida, 2004). The section was cut normal to the foliation and parallel to the lineation. Crossed polarized light. Scale bar: 3 mm. The amygdules are filled with calcite, which show many features of intracrystalline plasiticity viz. deformation twins, wavy extinction, and subgrain formation. The deformed amygdules have lenticular shapes due to precipitation and overgrowth of calcite grains in the extensional direction. *(Ichiko Shimizu)*

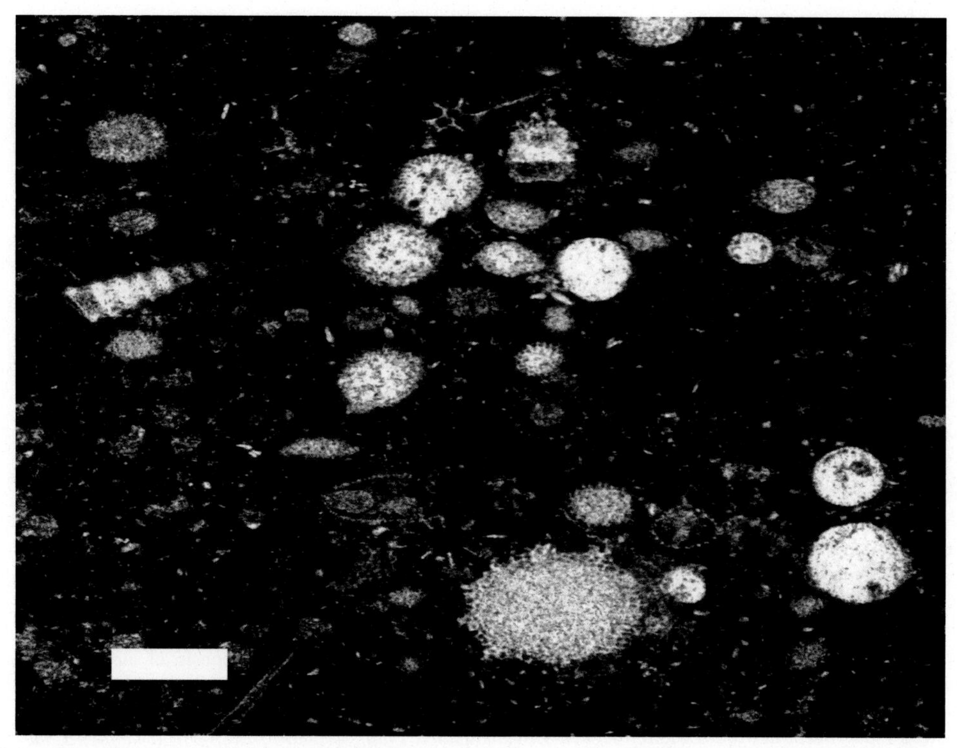

FIGURE 2.52 **Deformed radiolarian fossils in a black shale from a Jurassic accretionary complex (the Manba Unit) in the Northern Chichibu Belt, Central Japan (Shimizu, 1988).** The section was cut normal to the foliation. Plane polarized light. Scale bar: 200 μm. The shapes of initially spherical radiolarian fossils are direct indicator of strain ellipses. A corn-like radiolarian fossil (Nassellaria), hexagonal skeletons, and radial needles are also observed. Angular changes in these structures can also be used to quantify finite strain. *(Ichiko Shimizu)*

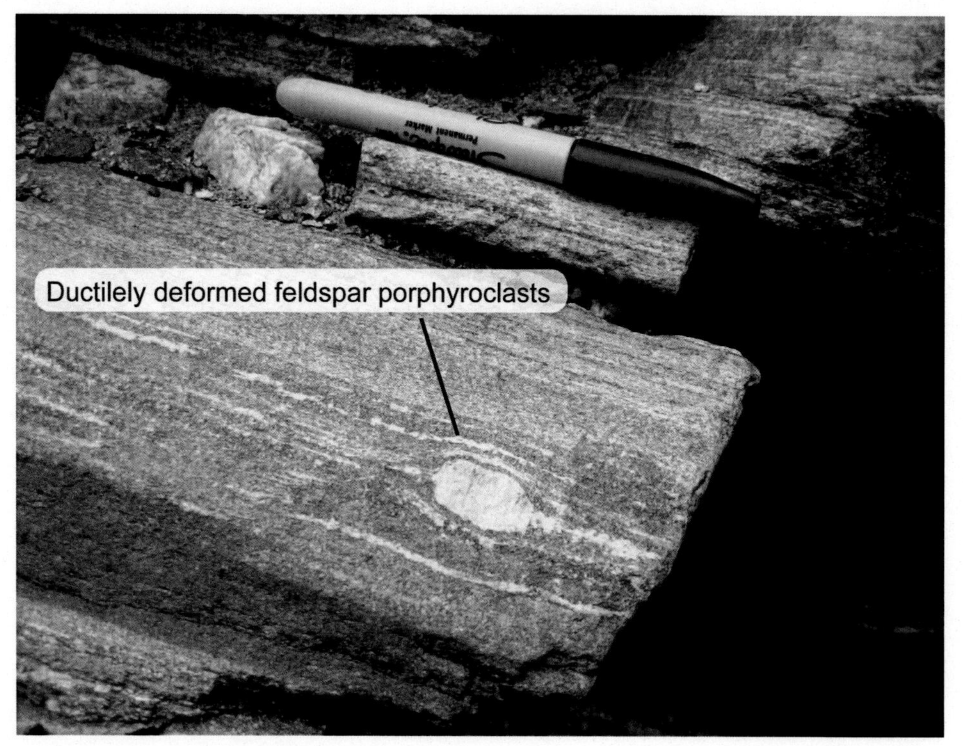

FIGURE 2.53 **Shear fabrics display a ductile deformed lenticular feldspar porphyroclast.** No shear sense indicated. Mylonitic foliation dips toward right. Ayishan detachment, West Tibet. *(Ran Zhang)*

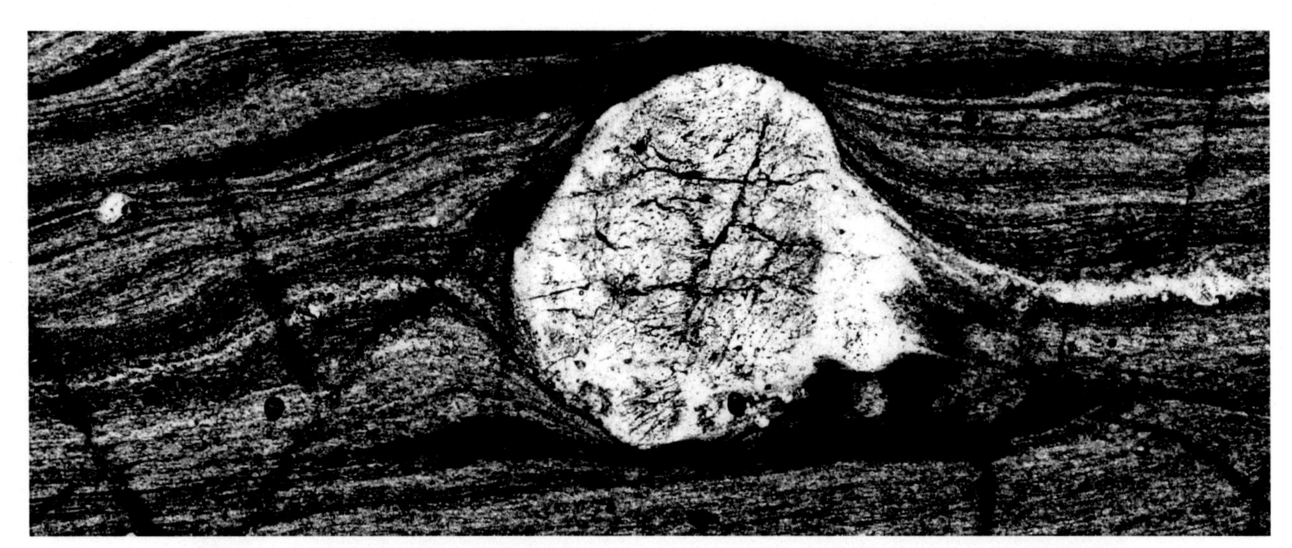

FIGURE 2.54 **Winged δ-microstructure of olivine.** The olivine grain deformed crystal-plastically under mantle condition in an ultramylonitized peridotite. The rocks crop out in a set of small islands in the Brazilian equatorial margins, the Archipelago Saint Peter and Saint Paul (ASPSP). The islands are one of the rare expositions of mantle rocks above the sea level. They are located along the active segment of the Saint Paul Transform Fault, in the Atlantic Mid Ocean ridge. The Olivine Forsterite is the main mineral in these rocks. In addition, clasts of orthopyroxene and spinel are found. The matrix consists of a very fine-grained crystal of forsterite and spinels. The microstructures consist of subgrains, deformation lamellae, and undulose extinction. Post-mylonitization fractures may also crosscut both clast and matrix and normally are filled with magnetite and serpentine. Adjacent to the clast, recrystallized olivine developed as asymmetric wings. Wing asymmetry indicates a top-to-left shear. Location: ASPSP (Brazil). Plane polarized light. Width of view 1.0 mm. *(Suellen Olívia Cândida Pinto, Leonardo Evangelista Lagoeiro, Luiz Sérgio Amarante Simões, Paola Ferreira Barbosa)*

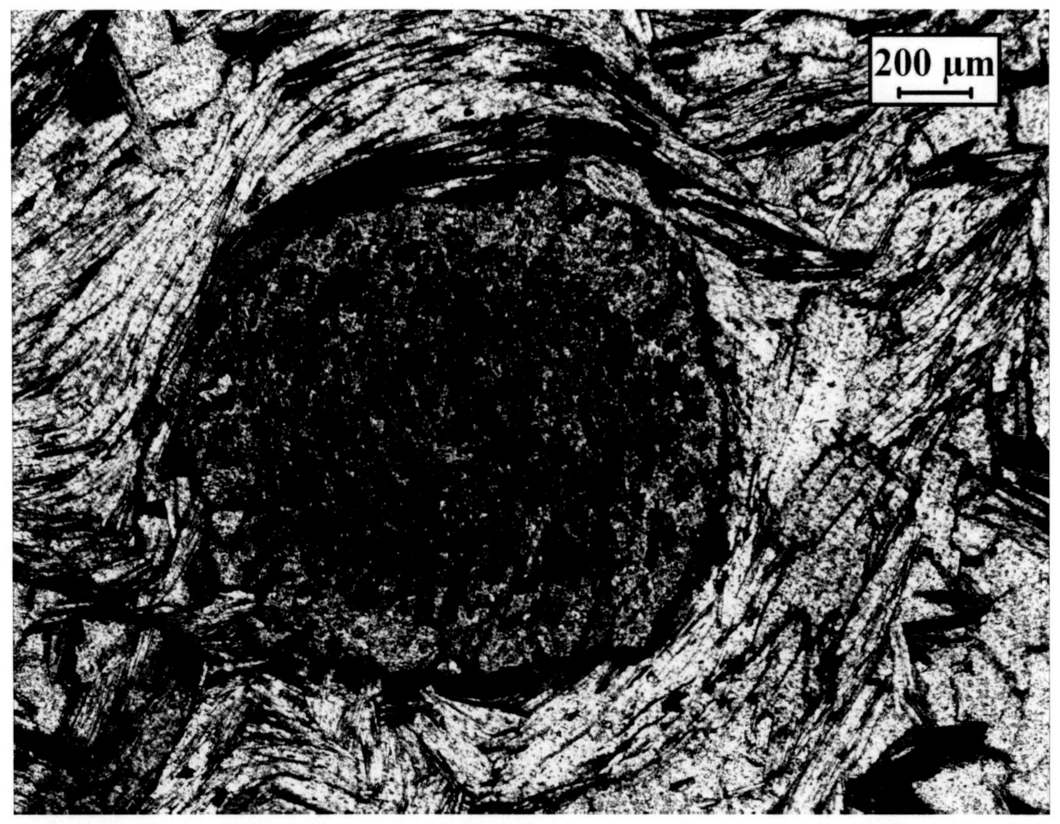

FIGURE 2.55 **Syntectonically grown garnet with sigmoid inclusion pattern.** Top-to-left sheared? Garnet-biotite schist of Proterozoic Daling Group. North to Gangtok, Sikkim, India. *(Atanu Mukherjee)*

REFERENCES

Bhadra, S., Gupta, S., Banerjee, M., 2004. Structural evolution across the Eastern Ghats Mobile Belt–Bastar Craton boundary, India: hot over cold thrusting in an ancient collision zone. Journal of Structural Geology 26, 233–245.

Bhadra S., Gupta S. Reworking of a basement-cover interface during terrane boundary shearing: an example from the Khariar basin, Bastar Craton, India. In: Mukherjee S., Mulchrone K.F. (Eds.), Ductile Shear Zones: From Micro- to Macro-scales. Wiley-Blackwell, in press.

Bikramaditya Singh, R.K., 2010. Geochemistry and mineral chemistry of granitoids of lesser himalayan crystallines, Western Arunachal Himalaya: implication for petrogenesis. Journal of the Geological Society of India 75, 618–631.

Biswal, T.K., Jena, S.K., Datta, S., Das, R., Khan, K., 2000. Deformation of the terrane boundary shear zone (Lakhna shear zone) between the Eastern Ghats Mobile Belt and the Bastar Craton, in the Balangir and Kalahandi districts of Orissa. Journal of the Geological Society of India 55, 367–380.

Biswal, T.K., DeWaele, B., Ahuja, H., 2007. Timing and dynamics of the juxtaposition of the Eastern Ghats Mobile Belt against the Bhandara Craton, India: a structural and zircon U–Pb SHRIMP study of the fold-thrust belt and associated nepheline syenite plutons. Tectonics 26, TC4006.

Clark, C., Fitzsimons, I.C.W., Healy, D., et al., 2011. How does the Continental crust gets really hot? Elements 7, 235–240.

Crowe, W.A., Nash, C.R., Harris, L.B., Leeming, P.M., Rankin, L.R., 2003. The geology of the Rengali Province: implications for the tectonic development of Northern Orissa, India. Journal of Asian Earth Sciences 21, 697–710.

Dabrowski, M., Graseman, B., 2014. Domino boudinage under layer-parallel simple shear. Journal of Structural Geology 68, 58–65.

Davis, G.H., Reynolds, S.J., Kluth, C.F., 2012. Structural Geology of Rocks and Regions. Wiley, New York.

Dobmeier, C., Raith, M., 2003. Crustal architecture and evolution of the Eastern Ghats Belt and adjacent regions of India. In: Yoshida, M., Windley, B.F., Dasgupta, S. (Eds.), Proterozoic East Gondwana: Supercontinent Assembly and Breakup, vol. 206. Geological Society of London, Special Publications, pp. 145–168.

Dutta, A., Gupta, S., Panigrahi, M.K., 2010. The southern Rengali province – a reworked or exotic terrane? Indian Journal of Geology 80, 81–96.

Engelder, T., Marshak, S., 1985. Disjunctive cleavage formed at shallow depths in sedimentary rocks. Journal of Structural Geology 7, 327–343.

Fagereng, A., Remitti, F., Sibson, R.H., 2010. Shear veins observed along planar anisotropy at high angles to greatest compressive stress. Nature Geoscience 3, 482–485.

Fagereng, A., Remitti, F., Sibson, R.H., 2011. Incrementally developed slickenfibres – geological record of repeating low stress-drop seismic events? Tectonophysics 510, 381–386.

Frehner, M., Exner, U., Mancktelow, N.S., Grujic, D., 2011. The not-so-simple effects of boundary conditions on models of simple shear. Geology 39, 719–722.

Fujimoto, Y., Yamamoto, M., 2010. On the granitoids in the Shirakami mountains and correlation to the Cretaceous to Paleogene granitoids distributed in the Northeast Japan. Earth Science 64, 127–144 (in Japanese with English abstract).

Ghosh, S.K., 1993. Structural Geology: Fundamentals and Modern Developments. Pergamon Press, Oxford. p. 598.

Ghosh, S.K., Hazra, S., Sengupta, S., 1999. Planar, non-planar and refolded sheath folds in Phulad shear zone, Rajasthan, India. Journal of Structural Geology 21, 1715–1729.

Grimmer, J.C., Glodny, J., Drüppel, K., Greiling, R.O., Kontny, A., 2015. Early- to mid-Silurian extrusion wedge tectonics in the central Scandinavian Caledonides Geology 43, 347–350.

Grimmer, J.C., Greiling, R.O., Gerdes, A., 2011. The Ammarnäs complex in the central Scandinavian Caledonides: an allochthonous basin fragment in the Sveconorwegian foreland? Terra Nova 23, 270–279.

ten Grotenhuis, S.M., Trouw, R.A.J., Passchier, C.W., 2003. Evolution of mica fish in mylonite rocks. Tectonophysics 372, 1–21.

Hada, S., Ito, M., Landis, C.A., Cawood, P., 2001. Large-scale translation of accreted terranes along continental margins. Gondwana Research 4, 628–629.

Harayama, S., Takahashi, Y., Nakano, S., Kariya, Y., Komazawa, M., 2000. Geology of the Tateyama District. With Geological Sheet Map at 1:50,000. Geological Survey of Japan. p. 218 (Japanese with English abstract 6 p).

Kontny, A., Engelmann, R., Grimmer, J.C., Greiling, R.O., Hirt, A., 2012. Magnetic fabric development in a highly anisotropic magnetite-bearing ductile shear zone (Seve Nappe Complex, Scandinavian Caledonides). International Journal of Earth Sciences 101, 671–692.

Koyi, H.A., Schmeling, H., Burchardt, S., Talbot, C., Mukherjee, S., Sjöström, H., 2013. Shear zones between rock units with no relative movement. Journal of Structural Geology 50, 82–90.

Lister, G.S., Snoke, A.W., 1984. S-C mylonites. Journal of Structural Geology 6, 617–638.

Mancktelow, N.S., Pennacchioni, G., 2005. The control of precursor brittle fracture and fluid-rock interaction on the development of single and paired ductile shear zones. Journal of Structural Geology 27, 645–661.

Mancktelow, N.S., Pennacchioni, G., 2013. Late magmatic healed fractures in granitoids and their influence on subsequent solid-state deformation. Journal of Structural Geology 57, 81–96.

Miklavič, B., Rožič, B., 2008. The onset of Maastrichtian basinal sedimentation on Mt. Matajur, NW Slovenia. RMZ – Materials and Geoenvironment 55, 199–214.

Mukherjee, S., 2007. Geodynamics, deformation and mathematical analysis of metamorphic belts of the NW Himalaya. Unpublished Ph.D. thesis Indian Institute of Technology Roorkee 1–267.

Mukherjee, S., 2010a. Structures at Meso- and Micro-scales in the Sutlej section of the Higher Himalayan Shear Zone in Himalaya. e-Terra 7, 1–27.

Mukherjee, S., 2010b. Microstructures of the Zanskar shear zone. Earth Science India 3, 9–27.

Mukherjee, S., 2011a. Flanking microstructures of the Zanskar shear zone, West Indian Himalaya. YES Bulletin 1, 21–29.

Mukherjee, S., 2011b. Mineral fish: their morphological classification, usefulness as shear sense indicators and genesis. International Journal of Earth Sciences 100, 1303–1314.

Mukherjee, S., 2012. Simple shear is not so simple! Kinematics and shear senses in Newtonian viscous simple shear zones. Geological Magazine 149, 819–826.

Mukherjee, S., 2013a. Deformation Microstructures in Rocks. Springer.

Mukherjee, S., 2013b. Higher Himalaya in the Bhagirathi section (NW Himalaya, India): its structures, backthrusts and extrusion mechanism by both channel flow and critical taper mechanism. International Journal of Earth Sciences 102, 1851–1870.

Mukherjee, S., 2013c. Channel flow extrusion model to constrain dynamic viscosity and Prandtl number of Higher Himalayan Shear Zone. International Journal of Earth Sciences 102, 1811–1835.

Mukherjee, S., 2014a. Atlas of Shear Zone Structures in Meso-scale. Springer.

Mukherjee, S., 2014b. Review of flanking structures in Meso- and Micro-scales. Geological Magazine 151, 957–974.

Mukherjee, S., 2014c. Mica inclusions inside host mica grains from the Sutlej section of the Higher Himalayan Crystallines, India: Morphology and Constrains in Genesis. Acta Geologica Sinica 88, 1729–1741.

Mukherjee, S., Biswas, R., 2014. Kinematics of horizontal simple shear zones of concentric arcs (Taylor–Couette flow) with incompressible Newtonian rheology. International Journal of Earth Sciences 103, 597–602.

Mukherjee S, Biswas, R. Biviscous horizontal simple shear zones of concentric arcs (Taylor Couette flow) with incompressible Newtonian rheology. In: Mukherjee S, Mulchrone K.F (Eds.), Ductile Shear Zones: From Micro- to Macro-scales. Wiley-Blackwell, in press.

Mukherjee, S., Koyi, H.A., 2009. Flanking microstructures. Geological Magazine 146, 517–526.

Mukherjee, S., Koyi, H.A., 2010a. Higher Himalayan Shear Zone, Sutlej section: structural geology and extrusion mechanism by various combination of simple shear, pure shear and channel flow in shifting modes. International Journal of Earth Sciences 99, 1267–1303.

Mukherjee, S., Koyi, H.A., 2010b. Higher Himalayan Shear Zone: Zanskar Indian Himalaya-Microstructural studies and extrusion mechanism by a combination of simple shear and channel flow. International Journal of Earth Sciences 99, 1267–1303.

Mukherjee, S., Mulchrone, K.F., 2013. Viscous dissipation pattern in incompressible Newtonian simple shear zones: an analytical model. International Journal of Earth Sciences 102, 1165–1170.

Mukherjee, S., Punekar, J.N., Mahadani, T., Mukherjee, R. Intrafolial folds-review & examples from the western Indian Higher Himalaya. In: Mukherjee S., Mulchrone K.F. (Eds.), Ductile Shear Zones: From Micro- to Macro-scales. Wiley-Blackwell, in press.

Mulchrone KF, Mukherjee S. Submitted. Kinematics and shear heat pattern of ductile simple shear zones with 'slip boundary condition'. International Journal of Earth Sciences.

Mulchrone KF, Mukherjee S. In press. Shear senses and viscous dissipation of layered ductile simple shear zones. Pure and Applied Geophysics. http://dx.doi.org/ 10.1007/s00024-015-1035-8.

Nabelek, P.I., Hofmeister, A.M., Whittington, A.G., 2011. The influence of temperature-dependent thermal diffusivity on the conductive cooling rates of plutons and temperature-time paths in contact aureoles. Earth and Planetary Science Letters 317, 157–164.

Pamplona, J., Rodrigues, B.C., 2011. Kinematic interpretation of shearband boudins: new parameters and ratios useful in HT simple shear zones. Journal of Structural Geology 33, 38–50.

Pamplona, J., Rodrigues, B.C., Fernandéz, C., 2014. Folding as precursor of asymmetric boudinage in shear zones affecting migmatitic terranes. Geogaceta 55, 15–18.

Passchier, C.W., Simpson, C., 1986. Porphyroclast systems as kinematic indicators. Journal of Structural Geology 8, 831–844.

Passchier, C.W., 2001. Flanking structures. Journal of Structural Geology 23, 951–962.

Passchier, C.W., Trouw, R., 2005. Microtectonics, Second ed. Springer, Berlin (Chapter 5, Shear zones, 336 p).

Pennacchioni, G., Mancktelow, N.S., 2007. Nucleation and initial growth of a shear zone network within compositionally and structurally heterogeneous granitoids under amphibolite facies conditions. Journal of Structural Geology 29, 1757–1780.

Platt, J.P., Vissers, R.L.M., 1980. Extensional structures in anisotropic rocks. Journal of Structural Geology 2, 397–410.

Ramsay, J.G., 1980. Shear zone geometry: a review. Journal of Structural Geology 2, 83–99.

Regenauer-Lieb, K., Yuen, D.A., 2003. Modeling shear zones in geological and planetary sciences: solid-and fluid-thermal-mechanical approaches. Earth- Science Reviews 63, 295–349.

Rutter, E.H., 1983. Pressure solution in nature, theory and experiment. Journal of the Geological Society, London 140 (5), 725–740.

Sengupta, S., Ghosh, S.K., 2004. Analysis of transpressional deformation from geometrical evolution of mesoscopic structures from phulad shear zone, Rajasthan, India. Journal of Structural Geology 26, 1961–1976.

Shimizu, I., 1988. Ductile deformation in the low-grade part of the Sanbagawa metamorphic belts in the northern Kanto Mountains, Central Japan. Journal of Geological Society of Japan 94, 609–628.

Shimizu, I., Yoshida, S., 2004. Strain geometries in the Sanbagawa metamorphic belt inferred from deformation structures in metabasite. Island Arc 13, 95–109.

Takahashi, Y., 2002. Granitic mylonites situated around the Shirakami Mountains, Northeast Japan. Earth Science 56, 215–216. In Japanese.

Takahashi, Y., Cho, D.L., Kee, W.S., 2010. Timing of mylonitization in the Funatsu Shear Zone within Hida Belt of southwest Japan: Implications for correlation with the shear zones around the Ogcheon Belt in the Korean Peninsula. Gondwana Research 17, 102–115.

Treagus, S., Lan, L., 2003. Simple shear of deformable square objects. Journal of Structural Geology 25, 1993–2005.

Trouw, R.A.J., Passchier, C.W., Wiersma, D.J., 2010. Atlas of Mylonites and Related Microstructures. Springer.

Vernon, R.H., 2004. A Practical Guide to Rock Microstructure. Cambridge University Press. 610 pp.

Chapter 3

Brittle Faults

KEYWORDS

Brittle shear zone; Brittle tectonics; Conjugate faults; Faults; Kinematic indicators; P-plane; Slickensides; Y-plane.

Brittle shear zones/fault zones are usually defined by curved brittle P-planes bound by usually straight Y-planes (Passchier and Trouw, 2005). These shears may affect as a narrow zone within the rock bodies (Misra et al. 2015). Brittle sheared lenses of rocks vary in geometry, and the P-planes may curve only near the Y-planes (Mukherjee, 2014a,b). Fault gouge zones sometimes contain P-planes that help to deduce the shear sense. Fault planes/Y-planes may contain slickensides. See Doblas (1998) for detail of slickenside types and their reliable use in shear sense determination. This is despite Tjia (1964) questioned reliability of slickensides as shear sense indicators. Deformational structures and especially faulted units within soft-sedimentary structures are quite common (Byrne, 1994) (Figures 3.1–3.49). In collisional tectonic regimes, brittle faults can form either in an in-sequence or in an out-of-sequence manner (Mukherjee, in press).

FIGURE 3.1 **Road-cut exposure of the Coal Creek fault zone, central eastern Front Range, Colorado, USA (39°54′14″N, 105°20′46″W).** This reverse fault zone occurs within the 1.7 Ga Boulder Creek, granodiorite batholith and hosts ~620-m west directed slip. The exposure shows a discrete, clay-rich gouge fault core and surrounding damage zone (cf. Caine et al., 1996). The damage zone is characterized by relatively low-temperature hydrothermal alteration concentrated mainly in hanging wall fractures juxtaposed against the pervasively and moderately altered clay-rich, white footwall. See Caine et al. (2010) for details. *(Jonathan Caine)*

Atlas of Structural Geology. http://dx.doi.org/10.1016/B978-0-12-420152-1.00003-X

FIGURE 3.2 **Fault core of the Coal Creek fault zone (See Figure 3.1).** Outcrop detail of the sharp, polished and striated principal slip surface hanging wall contact with the clay-rich gouge fault core. The gouge is composed of foliated, fault parallel white and red clay-rich gouge seams 1-cm to 3-cm thick. *(Jonathan Caine)*

FIGURE 3.3 **Polarized light micrograph of the red gouge from the Coal Creek fault zone (See Figure 3.2).** A relatively large quartz grain with asymmetric, granular "tails" show kinematic compatibility with the outcrop-scale indicators of contractional slip and is interpreted as evidence for cataclastic flow. The large central quartz grain also shows extensional fractures opened in a kinematically compatible direction and filled by calcite suggesting syntectonic fluid flow. The thin section is cut perpendicular to the plane of the fault and parallel to slip. *(Jonathan Caine)*

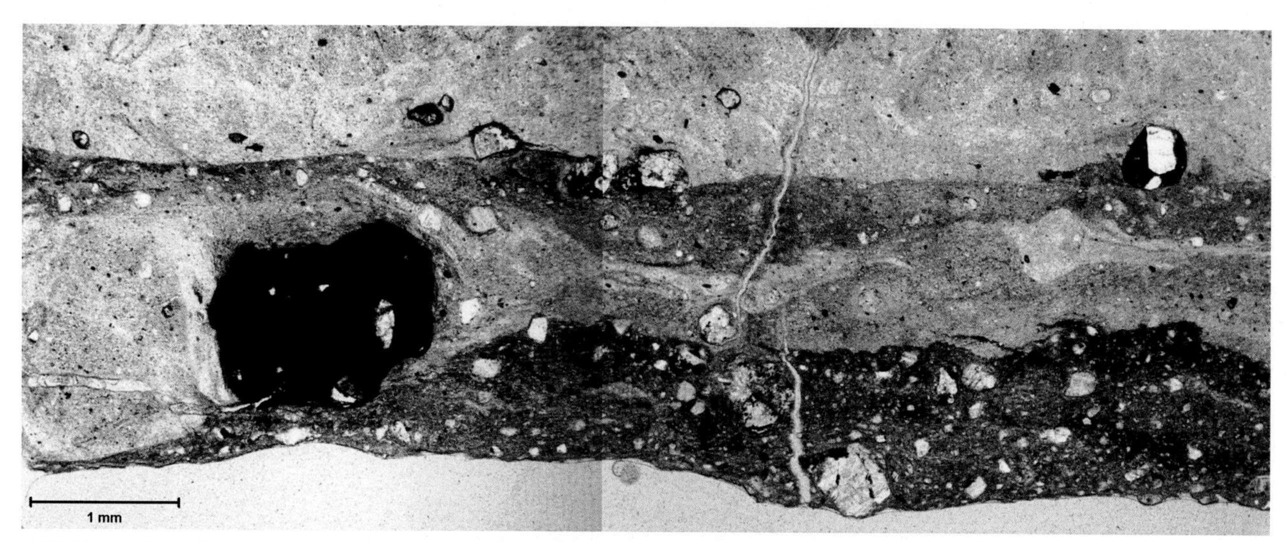

FIGURE 3.4 **Plain light micrograph of white and red clay-rich gouge from the Coal Creek fault zone (See Figure 3.2).** Clays envelop rounded quartz, feldspar, and lithic grains interpreted as evidence for episodic, cataclastic flow. The thin section is cut perpendicular to the plane of the fault and parallel to slip. *(Jonathan Caine)*

FIGURE 3.5 **Fault zone of the Raša fault.** Location: Senadole, Slovenia; WGS 84 coordinates: 45°43′00″ N, 14°00′15″ E; Rock type: bedded micritic limestone, Sežana formation, Late Cretaceous; Field of view: approx. 20 m. The Raša fault is a regional-scale seismically active fault with dextral to dextral-reverse kinematics inferred from fault-slip data and earthquake focal mechanisms (Poljak et al., 2000). A highway roadcut provides excellent exposure of its fault zone, which is here more than 100 m wide. The photograph shows the fault core (the right half of the picture) including fault gouge and the principal slip plane with well-developed slickensides (centre). Damage zone to the left is pervasively fractured and veined, to the extent that the originally sub-horizontal bedding is completely obliterated. *(Jonathan Caine)*

FIGURE 3.6 **Mesoscopic conjugate set of normal faults seen in a WNW-ESE section developed within thinly laminated sandstone-shale sequence of Early Cretaceous Lower Bhuj succession. Connotes extensional tectonic setting.** The fault planes strike NE-SW, and dip steeply ~70°. Slip is greater on the fault toward the right of the section, which displaces the earlier formed fault to the left. Right side of the photo is west direction. Width of the section: ~1.5 feet. On Kodki Road, Latitude 23°14′37.1″N, Longitude 69°34′59″E, near Bhuj city, Gujarat, India. *(Mainak Choudhuri)*

FIGURE 3.7 **Subvertical brittle shear Y-planes bound tangentially curved brittle P-planes, observed in a subvertical section.** Tip of pen indicates sense of shear. Psammitic schist. 24 19.903 N, 72 51.496 E. South Delhi Fold Belt. Near Rewari Dharmsala, Ambaji, Gujarat, India. See Mukherjee and Koyi (2010a,b), Mukherjee (2010a,b, 2012a,b, 2013a–c), Misra et al. (2014) etc. for more such features. *(Narayan Bose, Soumyajit Mukherjee)*

FIGURE 3.8 **A rare positive flower structure in the Precambrian quartzites (white) and metabasalts (black) in an abandoned quarry near Rajupalem, Andhra Pradesh, India.** The thrust verging toward W (white line) also shows a ramp-flat-ramp geometry. There are two dykes indicated by "D." Transpression probably during Pan-African orogeny (El-Wahed and Kamh, 2010). Width of view: ~150 m. Location: 14°0′37.96″N, 79°50′9.74″E. *(Achyuta Ayan Misra)*

FIGURE 3.9 **Curvilinear fault-bend folding of the Bhuj Sandstone at Lakhond in the Kachchh district of Gujarat, India.** The river that flows north has incised the exposure. Folding of the E-W striking beds is on account of the contraction (structural inversion) phase experienced by the Kachchh basin. *(George Mathew, Dnyanada Salvi)*

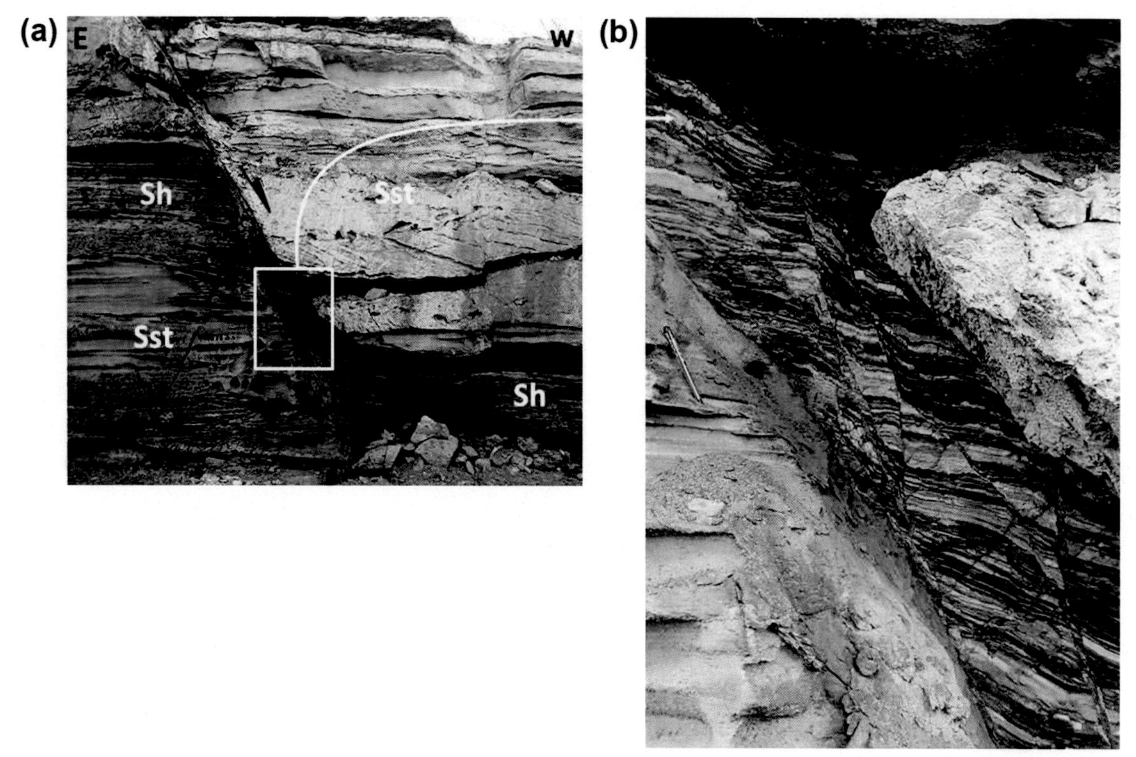

FIGURE 3.10 (a) A normal fault and (b) zoomed in showing the smear in the fault zone/overlap zone in a sand-shale sequence in the Cretaceous Bhuj Formation near Kodki village, E of Bhuj, Gujarat, India. The shale trapped and smeared in the overlap zone is characterized by smaller faults. Location: 23°14′37.59″N, 69°35′0.07″E. *(Achyuta Ayan Misra)*

FIGURE 3.11 **Antiformal stack duplex, Lesser Himalayan rocks exposed in the NE syntaxial zone of Arunachal Himalaya, India.** Arrows represent thrust slices.The exposed duplex lie within a smaller antiformal structure toward foreland bound at its N by Miri Thrust and S by the Main Boundary Thrust. The major crustal scale antiform, called the "Siang Antiform," bound this smaller antiform that is represented by the Gondwana Group of rocks. Location: 28°8′57.85″N, 95°13′23.41″E. Scale: Person (circled) height: 170cm. *(George Mathew, Dnyanada Salvi)*

FIGURE 3.12 **Uniform set of small-scale faults in a confined interval of parallel laminated lacustrine micrites.** Eocene Green River Formation, Fossil Basin, Wyoming, USA. Faults die out both upward and downward, and deformed interval bound by dark-brown (higher organic content), undisturbed laminated sediments. This indicates that deformation only affected rheologically susceptible sediments at the time. Slight changes in the thickness of the overlying sediments might indicate that deformation occurred before the deposition of the overlying sediments. The light color of the faulted sediments indicates low organic content, which might be responsible for contrasting lithological properties, and the formation of brittle deformation structures in the deposits. Faults might be related either to compaction-related extensional forces or seismic shocks. *(Reproduced from Figure 6 of Törő et al. (2013)–Balázs Törő, Brian R. Pratt, Sudipta Dasgupta)*

FIGURE 3.13 **A synsedimentary normal fault produced by gravitational effect.** Mudstone and graywacke. From the W side of the Rakhavdav complex, Udaipur district, Rajasthan, India. *(Moloy Sarkar)*

FIGURE 3.14 **Field photographs (a) and (b) demonstrate the range in thickness that can be observed in pseudotachylytes (PSTs).** Both images are of high-pressure (1.8–2.6 GPa) PST fault veins from the Schistes Lustres Complex, Cape Corse, Corsica (Austrheim and Andersen, 2004; Andersen and Austrheim, 2006; Ravna et al., 2010) (a) Shows an ultramafic PST vein approximately 35-cm thick hosted by peridotite wall rock. (b) Shows a much thinner fault vein (~2-mm thick) from the same area; it is hosted by a blueschist facies metagabbro wall rock (Andersen and Austrheim, 2006). Top-to-right shear sense displayed. *(Natalie Deseta)*

FIGURE 3.15 **Centimeter-scale top-to-NW brittle reverse fault within a quartzite clast.** The fault zone is ~1-cm thick. Mohand, Siwalik Himalaya, Roorkee–Dehradun transect, Uttarakhand, India. Srivastava and John (1999) reported the same shear sense from sandstones from this terrain. *(Tuhin Biswas)*

FIGURE 3.16 **Fascinating development of Y- and P-brittle shear planes within a single quartzite clast within sandy matrix.** Top-to-NNE (down) brittle shear indicated by small sigmoid P-planes. Near Mohand, Siwalik Himalaya, Roorkee–Dehradun transect, Uttarakhand, India. *(Tuhin Biswas)*

FIGURE 3.17 **Plan view of a sinistral strike-slip faulted pebble.** SSW geographical direction is toward right. 30°14.027′N, 77°56.871′E. Near Mohand, Roorkee–Dehradun transect, India. *(Dripta Dutta)*

FIGURE 3.18 **Subvertical fault plane cuts across the quartzite pebble within a conglomerate bed, Siwalik Supergroup.** Top-to-SW (up) shear. Tiny slip visible for the clast at the right side to the pen. Latitude: 30.2339°N, Longitude: 77.9438°E. Near Mohand, Roorkee–Dehradun transect, India. *(Dripta Dutta)*

FIGURE 3.19 **SE-dipping subparallel curved fault planes within a single quartzite pebble.** Siwalik Supergroup conglomerate. Long axis of the pebble: 14 cm. Near Mohand, Roorkee–Dehradun transect, Uttarakhand, India. *(Dripta Dutta)*

FIGURE 3.20 **A N-dipping subvertical fault plane cut across quartzite pebble.** Siwalik Supergroup conglomerate. Such a faulting indicates possibly an isostatic adjustment. Top-to-S (up) brittle shear. Near Mohand, Roorkee–Dehradun transect. 30°14.038′N, 77°56.9′E. Uttarakhand, India. *(Dripta Dutta)*

FIGURE 3.21 **Along-dip segmented normal fault and fault-related fold in Jurassic sequence.** Three fault segments dissect the condensed pelagic Jurassic sequence: one single planar plane in the lower part, two in the Tölgyhát Formation, one above the lower segment and one to the left (Sasvári et al., 2009). This disposition is typical for along-dip segmented normal faults (Childs et al., 1996; Rykkelid and Fossen, 2002). Between segments the deformation accommodated in the thin marlstone unit. The upper marlstone layers show folding between fault segments. The lower gray clay layers are boudinaged/pinched along the lower fault segment. Note the hanging wall part is present in the shadow of the marlstone layers. This different behavior could occur because of contrasting rheology of the deforming rocks. The segmented fault can have Middle Jurassic age (Bathonian, 168–166 Ma) because at least one upper segment is covered by Middle to Late Jurassic (Callovian–Oxfordian) radiolarite beds (Fodor et al., 2013). This early formation time matches with folding of the Toarcian unit, which deformed before complete diagenetic cementation, still in a semiplastic stage. The E-W-trending fault accommodated minor extension of a downbending side of a foreland basin opposite to the growing Dinaridic orogen (Fodor et al., 2013). Blue rucksack: ~50 cm. Early Jurassic limestone, late Early Jurassic (Toarcian) marlstone-claystone, Middle Jurassic nodular limestone of Tölgyhát Formation. Location: Tölgyhát quarry, Lábatlan village, Gerecse Hills, Hungary. Coordinates: 47°43′20.92″N, 18°30′45.80″E. *(László Fodor)*

FIGURE 3.22 **Top-to-SW brittle sheared or reverse faulted sandstone.** Deciphered from slip of brown bedding planes as markers. Near the pen marker, one more thinner marker set seen. Mohand, Siwalik Himalaya, Roorkee–Dehradun transect, Uttarakhand, India. *(Tuhin Biswas)*

FIGURE 3.23 **Kachchh Peninsula in Western India constitutes an active "fold-and-thrust" tectonic belt (Karanth and Gadhavi, 2007).** A curious structural feature is recorded in a road cutting near Nadapa (Latitude: 23°19′N Longitude: 69°52′E), about 5-km north of Bhuj, in a lithologically heterogeneous succession of a thick bed of sandstone overlain as well as underlain by thin beds of shale belonging to Jhuran Formation (Upper Jurassic). The pile of sediments subjected to compression exhibits development of folding and faulting in ductile shale layers on either side of the competent sandstone with opposite sense of vergence of folding and displacement (Figure 3.22). While the shale layers below the thick sandstone bed show northerly verging folds and north dipping fault planes (**Figure 3.23**); the ductile layers above sandstone reveal opposite sense of folding and faulting (**Figure 3.24**). Based on Teisseyre's analysis (1959) analogous movements have been suggested (reproduced in Figure 3X) by Jaroszewski (1984). Jaroszewski further strengthens his views quoting the results of the experiments conducted by B. Willis and R. Willis (reproduced in Figure 3Y). Along with several structural features such as low-angle reverse faults, fault-bend folds, bending moment faults, and drainage reversals, the present described feature supplements northerly directed compressive movements giving rise to "fold-and-thrust" tectonic belt in the region of Kachchh. Opposite sense of relative displacements in thin incompetent-shale layers on either side of a sandstone bed. *(R.V. Karanth, M.S. Gadhavi)*

FIGURE 3.24

FIGURE 3.25

FIGURE 3.26 **Imbricate thrusts possibly constitute an accretionary wedge.** Siwalik Himalayan sandstone of Dhokpathan Formation. Bagh Rao nala section Haridwar district, Uttarakhand, India. Outcrop Width ~200 cm. *(Shailendra Singh)*

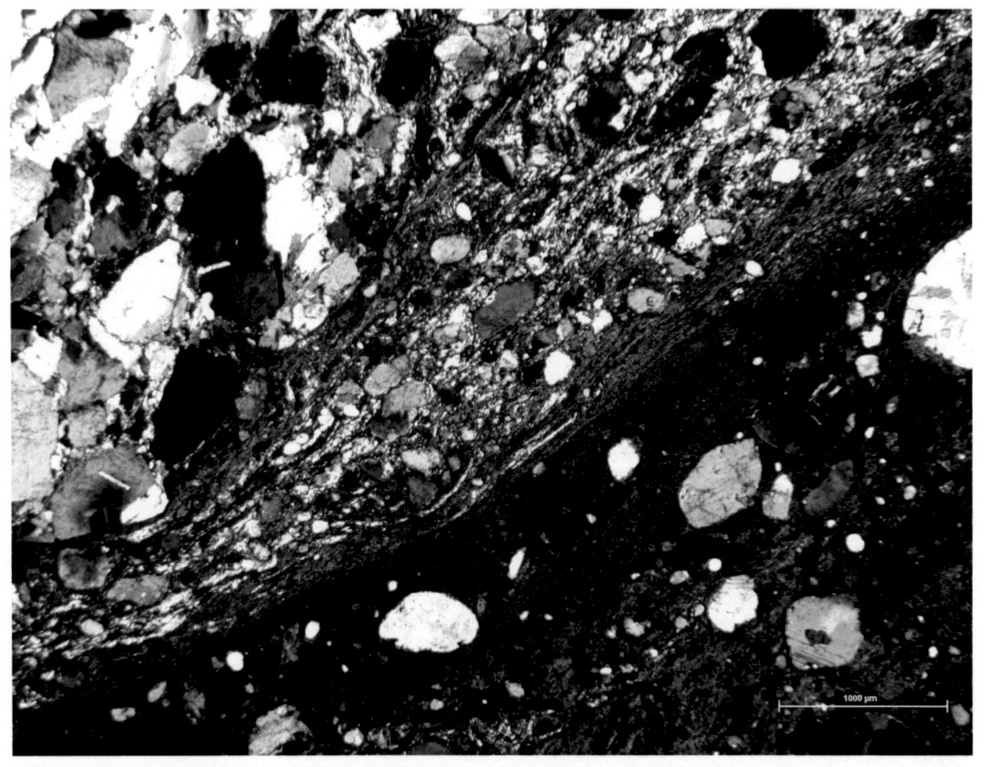

FIGURE 3.27 **Granodiorite of the Eagle Wash Intrusive Complex with mylonitic to cataclastic shear zones formed by rapid exhumation and cooling during detachment faulting** (Pease and Argent, 1999; Pease et al., 1999)**.** Note the fine-grained biotite matrix (lower right corner) grades into mylonitic ductile shear as the original igneous fabric of the granodiorite is approached at upper left. The dynamic recrystallization of quartz in the mylonitic shear zone is indicated by fine-grained ribbons, undulose extinction, parallel and elongate subgrains, and indicates >350 °C temperature (Simpson, 1985; Hirth and Tullis, 1994). Sacramento Mountains, California, Southwest US. Coordinates: N3853884, E710016. *(Victoria Pease)*

FIGURE 3.28 **Brittle tectonics is a powerful tool for studying rock characteristics in the field.** Results of a brittle deformation are tension or shear fractures and faults. On the base of fractures orientation and indicators of the movements along the planes, it is possible to determine the stress conditions when fractures (re)activated. Kinematic indicators on the fault surface decode slip sense (Doblas, 1998). The mineral accretion calcite steps (mineral fibers displayed on the fault plane in the figure) are commonly used slickenside indicators (Novakova, 2010). Here the mineral steps identify right lateral movement along the fault plane. From an abandoned quarry near the village Vapenna, Rychleby Mts., Czech Republic, Europe, Devonian crystalline limestones of Branna Group. GPS coordinates: N50°16′41″, E017°05′32″. *(Lucie Novakova)*

FIGURE 3.29 **A fault slickenside containing calcite slickenfibers.** Depending on conditions during brittle deformation, mineral fibers, e.g., calcite here, may grow at low-angle on rough fault planes on the facets that tend to open during slip. Mineral slickenfibers are asymmetric features that indicate slip sense: the congruous steps face toward the slip direction (Hancock, 1985; Petit, 1987). This picture evidences a synfolding fault slip, specifically (after Navabpour et al., 2007; with permission from Elsevier). The fault plane is subvertical and sinistrally offset the inclined strata. The plunge of striae is less than the dip of bedding plane, which is obvious at bottom left. This indicates that slip was synfolding after the strata tilted (Angelier, 1984). In this process, probably the fault plane also tilted. This is similar to a few flanking structure mechanism (see Mukherjee, 2014a,b). Picture spans ~15-cm width, from a NW-SE-trending fault plane in the Zagros Mountains, within Late Cretaceous marl of Gurpi Formation, W of Shiraz, Iran. *(Payman Navabpour)*

FIGURE 3.30 **A vertical fault with slickenside indicates oblique stylolites, also called slickolites.** Depending on the conditions in brittle deformation, soluble rock minerals, such as carbonates in this case, may dissolve from rough fault planes from facets involved with contraction during the fault slip to create oblique stylolites as pressure-solution seams. Oblique stylolites are asymmetric features that reveal slip sense, with the picks of stylolites at low-angle to the fault plane pointing the slip direction as incongruous steps (Hancock, 1985). This picture evidences a strike-slip fault slickenside with horizontal striae, specifically (personal observation). The fault plane is at right-angle to horizontal bedding plane of local strata and indicates a dextral slip. Picture spans ~15-cm width. Photographed by N. Kasch from a NNW-SSE-trending fault plane in the Thuringian Forest, within mid-Triassic shell-bearing limestone of the Muschelkalk lithostratigraphic unit, Germany. *(Payman Navabpour)*

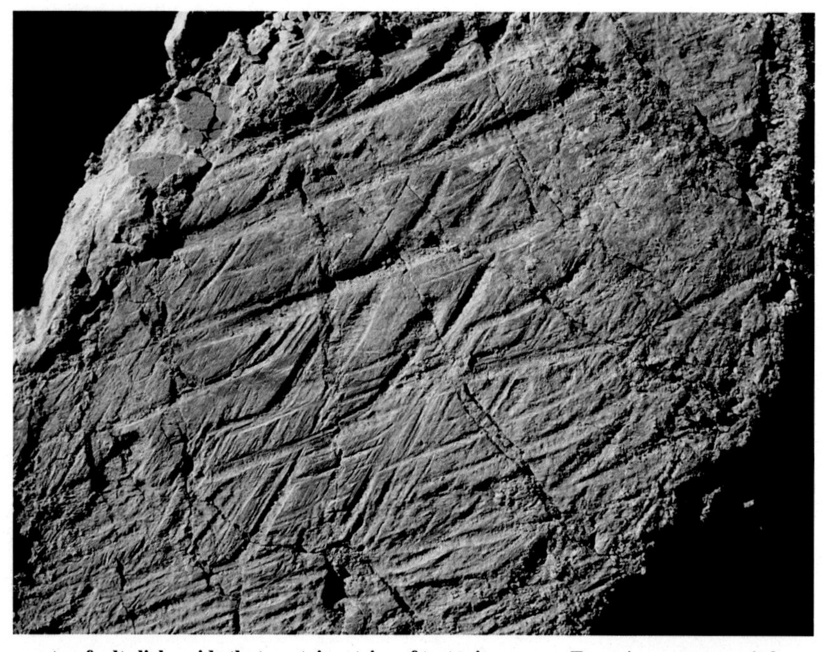

FIGURE 3.31 **Picture presents a fault slickenside that contains striae of tectonic grooves.** Tectonic grooves result from scratching of the fault plane by the opposite block. This creates linear tool marks. Depending on the conditions, tectonic grooves may indicate asymmetric geometries that can be used as kinematic indicators for slip sense (Hancock, 1985; Petit, 1987). This photograph shows crosscutting striae on a vertical fault slickenside (after Navabpour et al., 2007; with permission from Elsevier). The steeply dipping striae are subparallel to local bedding and are cut by the gently dipping striae, indicating a dextral shear. This feature evidences successive striae on a fault plane (Angelier, 1984), indicating a progressive counter-clockwise rotation around a horizontal axis perpendicular to the fault plane. At places, a gradual change from the first to the second striae is obvious. The picture spans ~15-cm width, and is taken from an E-W-trending fault plane in the Zagros Mountains, within Late Cretaceous marl of Gurpi Formation, NW of Shiraz, Iran. *(Payman Navabpour)*

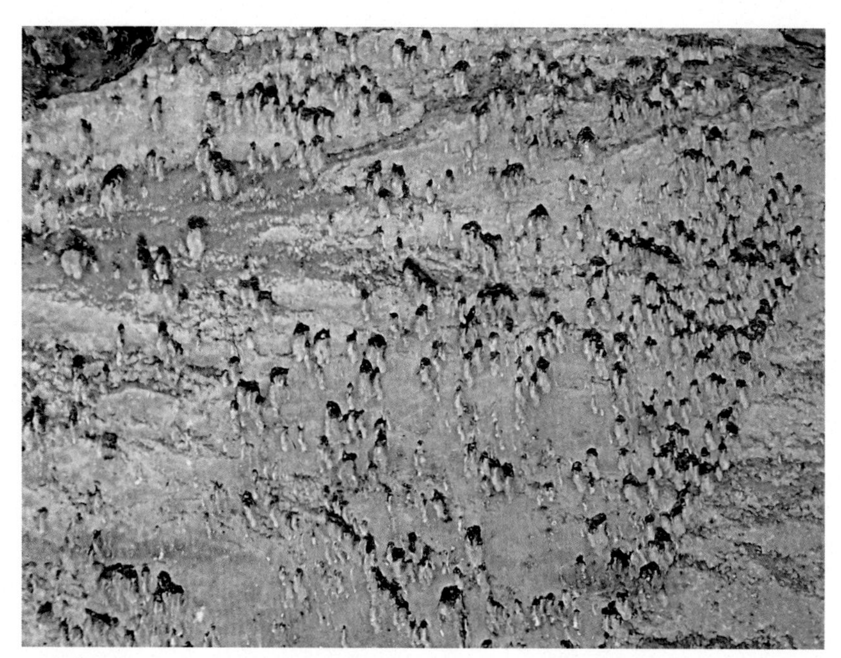

FIGURE 3.32 **A fault slickenside presents striae of erosive tectonic tools.** These tools are hard asperities of rock, such as garnet here, located between two faulted blocks or along a potential shear plane. During the slip on the fault plane, the asperities that are fixed within one block scratch the opposite block to create tectonic grooves, but at the same time they save rock materials behind themselves to leave linear rods in the direction of slip. As long as these asperities are preserved, the asymmetric geometry of the erosive tool marks can be used as a kinematic indicator for slip sense (Hancock, 1985; Petit, 1987). This picture indicates a normal fault slickenside, with downward linear rods of the saved rock materials behind the garnet grains. The picture spans ~15-cm width from a SE-dipping high-angle fault plane in the Zagros Mountains, within Late Cretaceous marl of Gurpi Formation, NW of Shiraz, Iran. *(Payman Navabpour)*

FIGURE 3.33 **Polished fault surface (slickenside or fault mirror), with well-developed striation which indicates vertical movement.** At the top and the left side of the image a brown fault gauge is also visible. Sandstone from a complex volcano-sedimentary terrain (former island arch). Sredna Gora Mountain, West-Central Bulgaria. *(Svetoslav Bontchev)*

FIGURE 3.34 **Slickensides on the fault plane of amphibolite gneiss.** These lineations are developed nonuniformly over the fault plane. Maithon, West Bengal, India. *(Ananya Basu)*

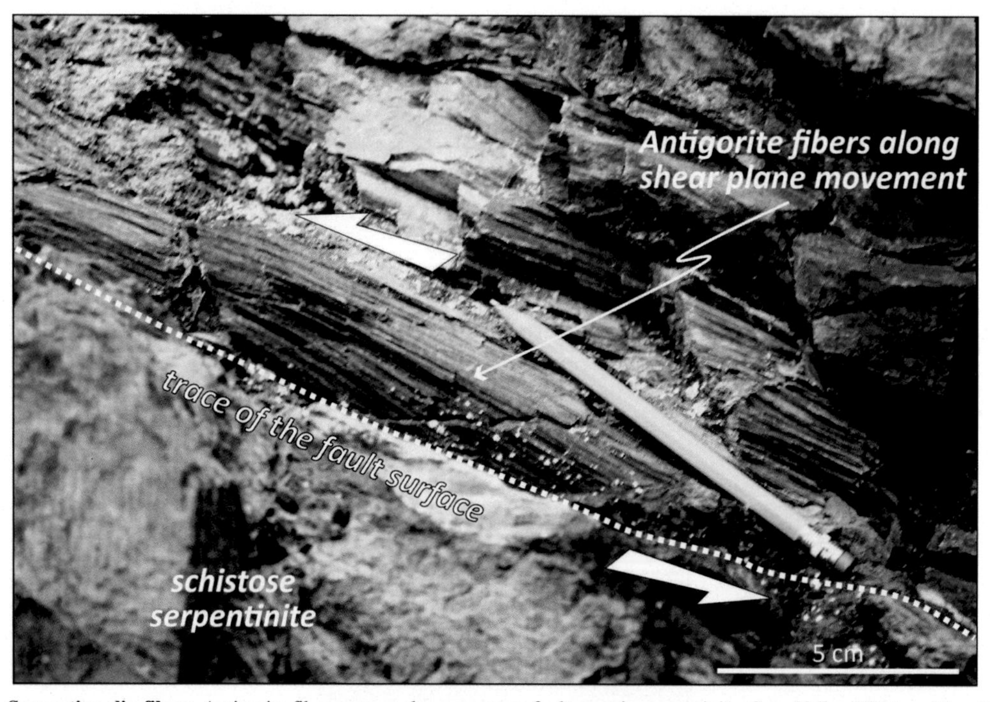

FIGURE 3.35 **Serpentine slip fibers.** Antigorite fibers grown along a reverse fault zone in serpentinite, Susa Valley, Western Alps, Italy. The fibrous morphology (i.e., the length is at least three times the width) and crystallization of antigorite are due to fluid-assisted shearing. The fibrous minerals are disposed parallel to the shear surface and can be interpreted as slip-fibers (Ramsay and Huber, 1983). The studied rocks are part of the Liguro-Piemontese oceanic domain of the Western Alps (Polino et al., 1990). These ophiolitic units have been involved in the subduction/exhumation cycle during orogenic construction (Agard et al., 2009) and they are affected by ductile-to-brittle deformation fabrics, locally hosting fibrous mineralization (e.g., Compagnoni and Groppo, 2006). *(Gianluca Vignaroli, Federico Rossetti)*

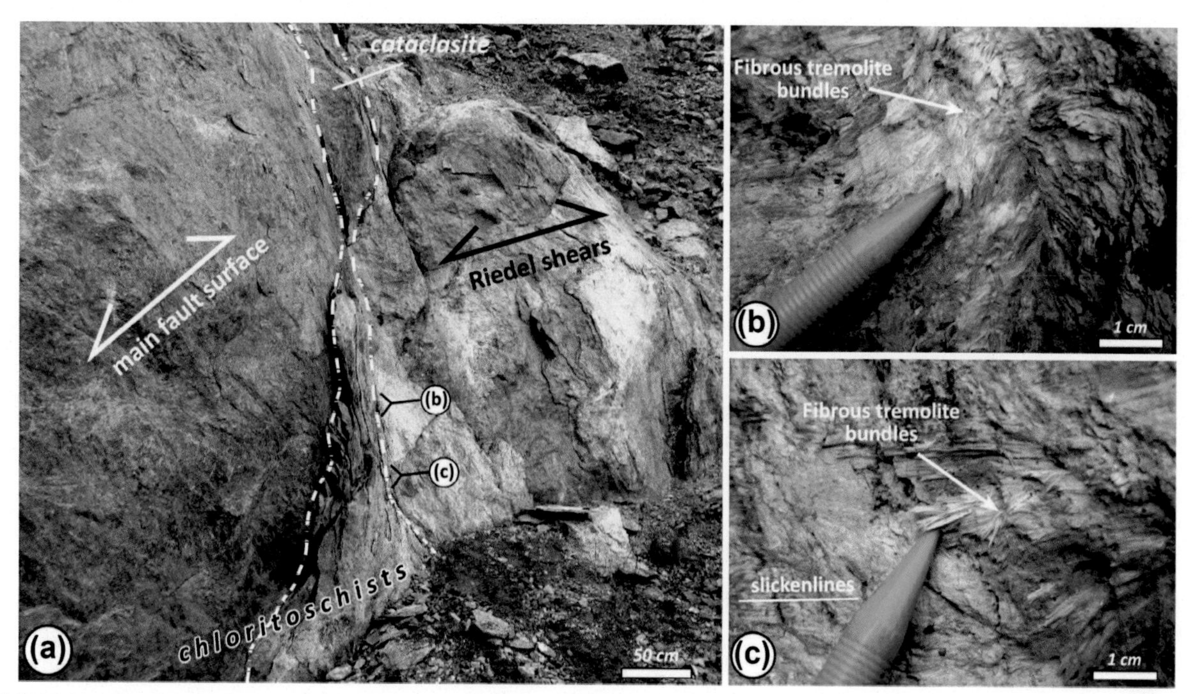

FIGURE 3.36 The structural control on asbestos mineralization in mafic rocks. (a) Chloritoschists affected by strike-slip faulting, Voltri Massif (Ligurian Alps, Italy). The fault zone consists of a decimeter-thick cataclastic band parallel to the main fault surface. Synthetic Riedel shear planes intersect at low-angle the main fault surface and attest dextral kinematics (e.g., Petit, 1987). Asbestos mineralization occurs within the fault damage zone, localized where the main fault interferes with the Riedel shears. (b), (c) Detail of asbestos mineralization that decorates the Riedel shears. Asbestos mineralization consists of up to 2-cm long tremolite fibers, commonly elongated parallel to the fault slickenlines. The Voltri Massif is an ophiolitic domain located at the Alps-Apennines junction in Northern Italy. It includes subduction-related exhumed high-pressure (variably retrogressed blueschist-to-eclogite) units (Vanossi et al., 1984). During the exhumation of the high-pressure units, ductile-to-brittle deformation structures developed, and hosted composite mineralogical assemblages associated with retrogressive metamorphism (Federico et al., 2007; Vignaroli et al., 2010). Fibrous mineralization usually occurs in semi-brittle and brittle deformation structures. *(Gianluca Vignaroli, Federico Rossetti)*

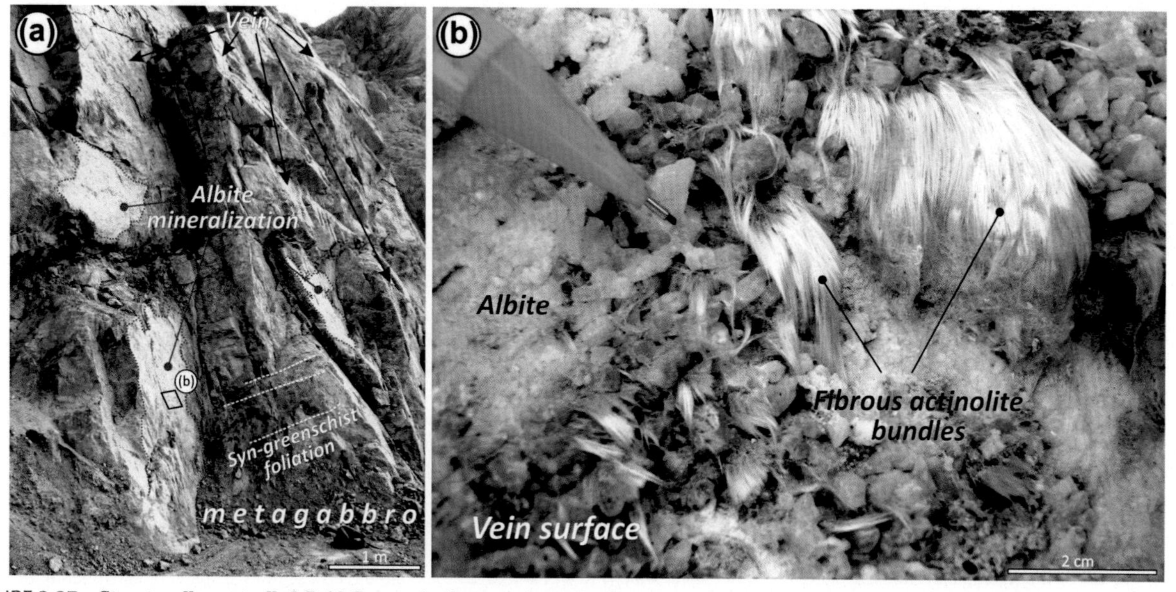

FIGURE 3.37 Structurally controlled fluid-flow and asbestos mineralization. (a) Sub-vertical vein system that cuts across the syn-greenschist foliation in metagabbros, Voltri Massif (Ligurian Alps, Italy). The veins are made up by albite with subordinate actinolite amphibole. (b) Detail of the vein surface where fibrous actinolite crystallizes. Fibers growth occurs in square centimeter bundles. This example documents percolation of mineralizing fluids assisted by structurally controlled permeability network. Metamorphic veins with fibrous minerals are described commonly in ophiolitic rocks (Hoogerduijn Strating and Vissers, 1994; Karkanas, 1995; Andreani et al., 2004). These tensional fractures can be thought as dilatation sites where fibrous mineral form along with fluid circulation and confined chemical mass transfer within the vein width (Barker et al., 2006). For references on the regional geology, see the previous caption. *(Gianluca Vignaroli, Federico Rossetti)*

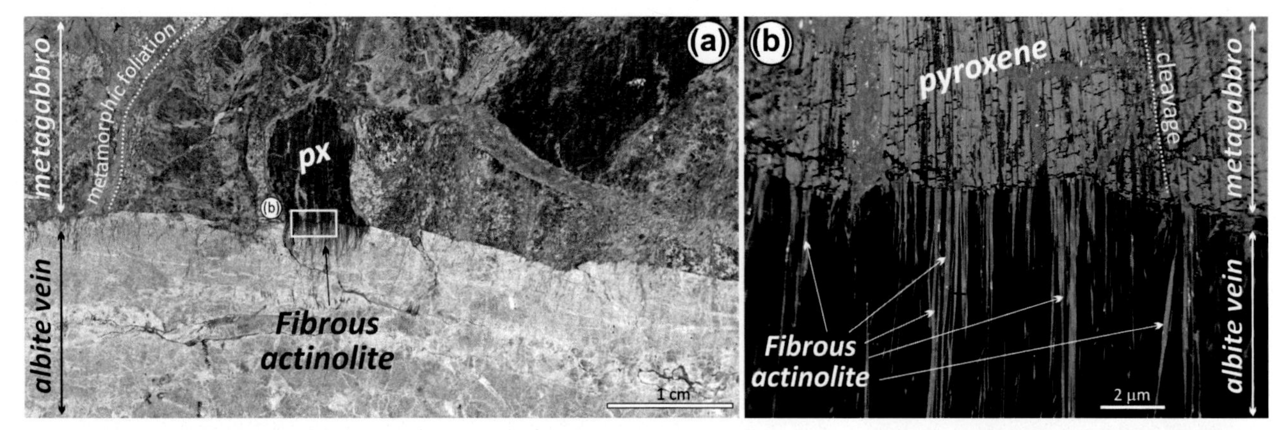

FIGURE 3.38 **The structural control on asbestos mineralization at microscale.** (a) Detail of the sharp boundary between the metagabbro host rock and the albite-actinolite vein shown in Figure 3.37. The vein shows an antitaxial growth (Passchier and Trouw, 2005). The pyroxene (px) grain from the metagabbro is truncated at the vein-wall interface. Half a centimeter long fibers of actinolite concentrate along the pyroxene-vein interface. (b) This back-scattered electron image is a detail of the textural relationships between pyroxene and fibrous actinolite. Fibers occurrence parallel to the internal cleavage of the pyroxene suggests a textural control of the pyroxene lattice during the actinolite overgrowth. Fibrous actinolite is one of the six asbestos minerals regulated by the normative (World Health Organization, 1986). The asbestos hazard in natural environment is connected to the rock fabric heterogeneities (ductile-to-brittle deformation localization, metamorphism/metasomatism, and structurally controlled fluid flow) developed in response to the regional tectonics (Vignaroli et al., 2011). *(Gianluca Vignaroli, Federico Rossetti)*

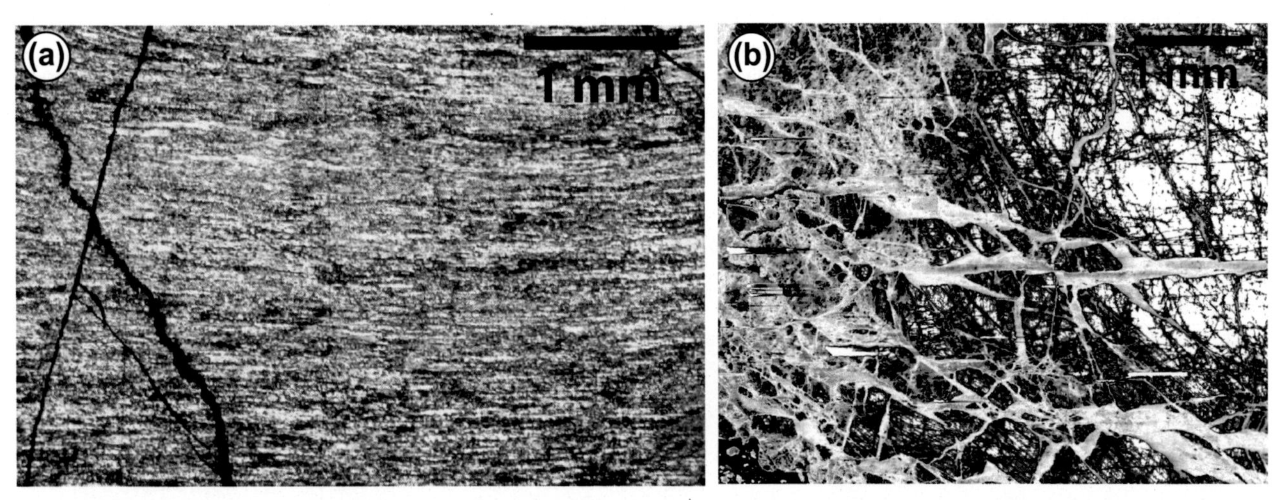

FIGURE 3.39 & FIGURE 3.40 **Microstructural evidence for repeated cycles of coseismic damage and viscous flow at the frictional to viscous transition in an ancient strike-slip fault.** Large-displacement continental strike-slip faults are characterized by a core of ultrafine-grained rocks surrounded by a "damage-zone" of fractured rocks that can extend laterally outward for hundreds of meters. A notable characteristic of these faults is the extensive occurrences of the so-called "pulverized" rocks that exhibit intragranular cracking and fragmentation down to the micron scale. A fundamental question is how deep do these damage zones extend into the crust, and how do their physical characteristics change with depth (Mooney et al., 2007; Tullis et al., 2007)? The Paleozoic Norumbega Fault System in Maine, USA, is known to be one of the best ancient analogs for the San Andreas Fault currently exposed on Earth's surface (Ludman and West, 1999; Sibson and Toy, 2006). The Norumbega is exposed at a range of depths along its length, and in Central Maine the mylonites along different fault strands preserve spectacular evidence for repeated coseismic brittle damage followed by viscous flow, representing the frictional to viscous transition. We conclude from our studies of these rocks that damage zones extend throughout the seismogenic zone, but are difficult to detect geophysically and also by optical methods at depth owing to rapid microcrack healing and interseismic viscous flow. Some microstructural evidence is visible optically, but most of it can only be observed using cathodoluminescence (CL) and electron backscatter diffraction analyses. The following figure pairs from the Sandhill Corner shear zone (Price et al., 2012) illustrate the efficacy of CL imaging in extracting detailed microstructural history that is not visible optically. Optical (a: cross-polarized light) and CL (b) images of a shattered plagioclase grain. Optically, the grain shows at least three generations of crosscutting quartz-filled cracks, and patchy extinction. The CL image reveals remarkable complexity not visible optically. The dark patches are extensive tensile microcrack networks in plagioclase. A dense network of quartz-filled cracks overprint them. Plagioclase fragments are as fine as 1 μm. *(Scott Johnson)*

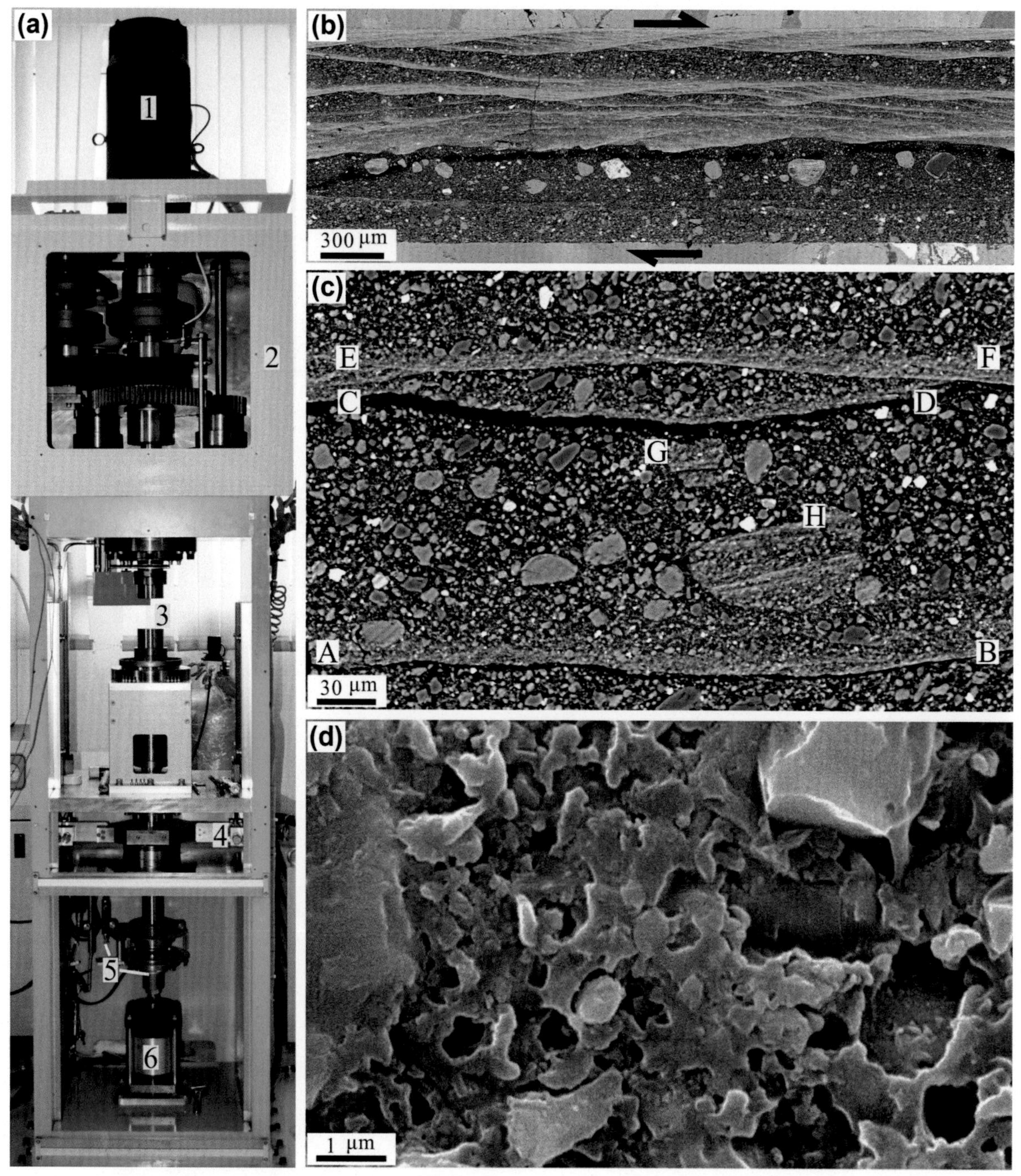

FIGURE 3.41 **Overlapped slip-zone structures in experimental fault gouge.** Reproducing seismic fault motion in laboratory received attention in the last two decades. Our institute installed a low to high velocity friction apparatus (a) that produces plate to seismic velocities (60 mm yr^{-1} to a few m s^{-1}). The apparatus consists of (1) servomotor, (2) gear/belt system for changing velocity, (3) holders of a pair of cylindrical specimens, (4) torque gauge, (5) axial-force and displacement transducers, and (6) bellow cylinder for applying the axial force. Sometimes very complex slip-zone structures such as those formed in experimentally deformed gouge have been noted. (b): Gouge was collected from the Pingxi fault zone in the Longmenshan fault system that caused the 2008 Wenchuan earthquake and was deformed with room humidity at a slip rate v of 1.4 m s^{-1} and a normal stress σ_N of 0.8 MPa (Yao et al., 2013). The gouge consists of highly deformed slip zones, that appear white on the photograph and weakly deformed gouge. Crosscutting relations are recognized clearly along Riedel shears within gouge. Many of those slip zones are characterized by asymmetric grain-size distributions with ultrafine-grains at the bottom coarsens gradually upward (see three slipping zones AB, CF and EF in (c) from the same gouge.) The middle slip zone is truncated by the upper slip zone at point F, and two fragments of slip zones are scattered at points G and H. Ultrafine-grained zones often consist of sintered grains with some pores (e.g., quartz gouge in (d) deformed at v = 1.3 m s^{-1} and σ_N = 3.1 MPa: Togo and Shimamoto 2012). Highly sheared gouge is strong, and slip zone can shift to less-deformed weaker gouge once the gouge is welded or hardened by compaction. This develops complex slip-zone structures. We suggested that welding or compaction of gouge can be a mechanism for the growth of gouge zone in natural fault zones (Shimamoto and Togo, 2012). *(Shengli Ma, Lu Yao, Tetsuhiro Togo, Toshihiko Shimamoto)*

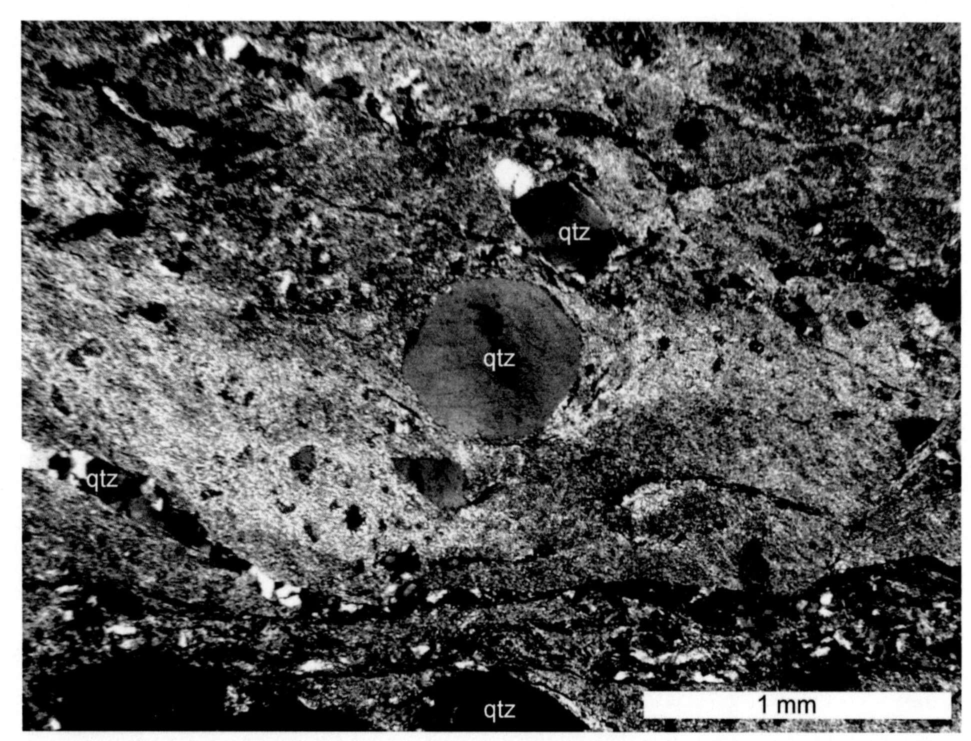

FIGURE 3.42 **Sigma-structure in ultracataclastic, fine-grained matrix of quartz and clay around quartz-clast indicates foreland-directed (top-to-right) tectonic transport.** Lower Allochthon, Central Scandinavian Caledonides. See Greiling et al. (1998) for detail. *(Jens Carsten Grimmer)*

FIGURE 3.43 **Brittle-ductile thrust shear zone related to the NNE-SSW-trending Olevano-Antrodoco-Sibillini oblique thrust ramp, Central-Northern Apennines, Italy.** The foliated cataclasite is organized in a planar S-fabric (Calamita et al., 2012). The penetrative subhorizontal closely-spaced pressure-solution S-cleavage affects the marly-calcareous Eocene–Oligocene Scaglia Cinerea Formation cropping out in the footwall of the thrust ramp. Conjugate high-angle extensional shear planes occur within the thrust shear zone and are related to extensional crenulation cleavage as per Calamita et al. (2012). Details about the brittle-ductile thrust shear zones in the Central-Northern Apennines can be found in Calamita (1991), Tavarnelli (1999), and Calamita et al. (2012). Location: Valle Scura Valley (42.490855° 13.049969°), W to the Sigillo village, Province of Rieti (Italy). Width of view: 70 cm. *(Paolo Pace)*

FIGURE 3.44 **Synlithification faults in Cretaceous clastics.** In the gray marl, the centimeter-thick sandstone layers commonly contain a set of small, parallel displacement zones, which cannot be followed up to the next sandstone bed. Within the sandstone, the fault planes are frequently not discrete surfaces and are macroscopically invisible. They either do not continue in the intercalating marlstone or occur as closed fractures. In addition, displacement lines at the upper and lower boundaries of the sandstone bed are not always aligned but are en-echelon along dip direction. This can be interpreted as along-dip segmented fractures. Displacement can be smaller at the upper than at the lower bedding plane. All these deformation features developed when sandstone beds consolidated/cemented partly and were in plastic state so they are synlithification faults. The deformation can be regarded as a first step for boudinage. Deformation occurred during the Early Cretaceous progressive burial. Compaction affected the conjugate fracture set and increased their angle to obtuse angle. Postdeformation cementation sealed the early displacement features. The deformation is related to an Early Cretaceous foreland basin formation (Tari, 1994; Fodor et al., 2013). Green eraser: ~ 2 cm. Early Cretaceous (Valanginian) marlstone, sandstone (Fogarasi, 1995). Location: Bersek quarry, Lábatlan village, Gerecse Hills, Hungary. Coordinates: 47°43′20.33″N, 18°31′29.13″E. *(László Fodor)*

FIGURE 3.45 **The outcrop exposes one of the main faults of the active Hronov-Poříčí Fault Zone, Czech Republic.** It is ~70-km long intraplate zone responsible for earthquakes up to M~4.7 recently (Woldrich, 1901; Špaček et al., 2006). Red-brown Permian conglomerates and breccias of the Trutnov Formation on the right side form hanging wall of the fault zone. Conglomerates are overlain unconformably by ochre to dark gray Cretaceous sandstones and siltstones of the Peruc-Korycany Formation, cropping out on left side of the wall (see Novakova, 2014). The dip of both the Permian and the Cretaceous sediments is steep (up to 89° SW) due to the close vicinity of the main fault line. The Hronov-Poříčí Fault Zone represents main reverse fault accompanied by parallel/oblique normal or reverse faults (Valenta et al., 2011). GPS coordinates: N50°32′14″, E016°02′31″. *(Lucie Novakova)*

FIGURE 3.46 **This view from Point Keen above Kaikoura (South Island, New Zealand) displays a set of anastomosed sinistral faults.** These relatively small faults are related to the main Hope Fault. The Hope Fault is currently the most active structure of the Marlborough Fault System with a right-lateral slip rate of 20–40 mm/yr (Cowan, 1991). The Northern Canterbury basin and range topography is a consequence of spreading deformation associated with the Australian-Pacific plate boundary (Rattenbury et al., 2006). GPS coordinates: S42°25′26″, E173°43′01″. *(Lucie Novakova)*

FIGURE 3.47 **Multiple generations of pseudotachylyte.** This back-scattered electron image shows an Eocene blueschist facies, peridotite-hosted pseudotachylyte (PST). At least three generations of faulting whereby the earliest generation of pseudotachylyte (PST 1) has been cross-cut by two later generations (PST 2 and PST 3; Sibson, 1975; Spray, 1992; Swanson, 1992). PST 2 is microfaulted implying the PST had time to solidify between pseudotachylyte melt generations. The dark grey PST 3 remains glassy and has entrained chunks of the previous PST generations. This PST is situated in the exhumed ophiolite complex: Schistes Lustres, Cape Corse, Corsica (Andersen and Austrheim 2006). *(Natalie Deseta)*

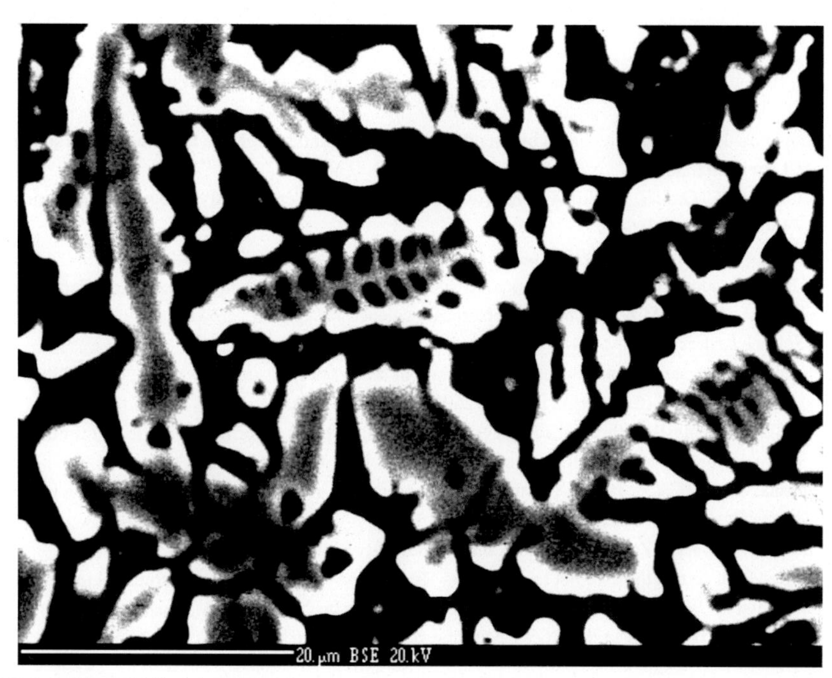

FIGURE 3.48 **High-Al Omphacite.** The first documented occurrence of omphacite hopper crystals quenching directly from a pseudotachylytic melt, suggesting high pressure (1.8–2.6 GPa), high temperature (>1350 °C) conditions of formation (Andersen and Austrheim, 2006; Ravna et al., 2010). These mafic pseudotachylytes are hosted by blueschist-eclogite-lawsonite facies metagabbro from the exhumed Schistes Lustres Ophiolite Complex, Cape Corse, Corsica. These omphacites contain an average of 15 wt% Al_2O_3 and a ~20 mol% Ca-Eskola component (Smyth, 1980; Katayama et al., 2000; Deseta et al., 2014). This high temperature occurrence of omphacite is important as it confirms that these pseudotachylytes formed under high pressure conditions. *(Natalie Deseta)*

FIGURE 3.49 **Imbricate thrust structure in flysch.** This spectacularly exposed structure is located in the most external part of the Dinaric fold-thrust belt in the Istria peninsula (Placer et al., 2010). The floor detachment runs ~1-m above prominent sandstone beds at the foot of the cliff. The isolated position of this duplex within the undisturbed flysch sequence, and markedly plastic behavior of sandstone layers during deformation, indicated by tight drag folds in footwall of imbricate thrusts, suggest that the structure might have originated due to soft-sediment slumping rather than tectonic deformation. Rock type: thin-bedded sandstone, marlstone, and mudstone, Dinaric foreland flysch, Eocene. Location: San Simon bay, Adriatic coast, Slovenia. Coordinates: 45°32′00″N, 13°38′25″E. *(Marko Vrabec)*

REFERENCES

Agard, P., Yamato, P., Jolivet, L., Burov, E., 2009. Exhumation of oceanic blueschists and eclogites in subduction zones: timing and mechanisms. Earth-Science Reviews 92, 53–79.

Andersen, T., Austrheim, H., 2006. Fossil earthquakes recorded by pseudotachylytes in mantle peridotite from the Alpine subduction complex of Corsica. Earth and Planetary Science Letters 242, 58–72.

Andreani, M., Baronnet, A., Boullier, A.M., Gratier, J.P., 2004. A microstructural study of a "crack-seal" type serpentine vein using SEM and TEM techniques. European Journal of Mineralogy 16, 585–595.

Angelier, J., 1984. Tectonic analyses of fault slip data sets. Journal of Geophysical Research 89, 5835–5848.

Austrheim, H., Andersen, T.B., 2004. Pseudotachylytes from Corsica: fossil earthquakes from a subduction complex. Terra Nova 16, 193–197.

Barker, S.L.L., Cox, S.F., Eggins, S.M., Gagan, M.K., 2006. Microchemical evidence for episodic growth of antitaxial veins during fracture-controlled fluid flow. Earth Planetary Science Letters 250, 331–344.

Byrne, T., 1994. Sediment deformation, dewatering and diagenesis: illustrations from selected mélange zones. In: Maltman, A. (Ed.), The Geological Deformation of Sediments. Chapman & Hall (Chapter 8), pp. 239–260.

Calamita, F., Satolli, S., Turtù, A., 2012. Analysis of thrust shear zones in curve-shaped belts: deformation mode and timing of the Olevano-Antrodoco-Sibillini thrust (Central/Northern Apennines of Italy). Journal of Structural Geology 44, 179–187.

Calamita, F., Decandia, F.A., Deiana, G., Fiori, A.P., 1991. Deformation of s-c tectonites in the scaglia cinerea formation in the Spoleto area (South-East Umbria). Bollettino della Società Geologica Italiana 110, 661–665.

Caine, J.S., Evans, J.P., Forster, C.B., 1996. Fault Zone Architecture and Permeability Structure. Geology 24, 1025–1028.

Caine, J.S., Ridley, J., Wessel, Z.R., 2010. To reactivate or not to reactivate—Nature and varied behavior of structural inheritance in the proterozoic basement of the eastern Colorado Mineral Belt over 1.7 billion years of earth history. In: Morgan, L.A., Quane, S.L. (Eds.), Through the Generations: Geologic and Anthropogenic Field Excursions in the Rocky Mountains from Modern to Ancient. Geological Society of America Field Guide, vol. 18, pp. 119–140.

Childs, C., Nicol, A., Walsh, J.J., Watterson, J., 1996. The growth of vertically segmented normal faults. Journal of Structural Geology 18, 1389–1397.

Compagnoni, R., Groppo, C., 2006. Gli amianti in Val di Susa e le rocce che li contengono. Rendiconti della Società Geologica Italiana 3, 21–28.

Cowan, H.A., 1991. The North Canterbury earthquake of September 1, 1888. Journal of the Royal Society of New Zealand 21, 13–24.

Deseta, N., Andersen, T.B., Ashwal, L., 2014. A weakening mechanism for intermediate-depth seismicity? Detailed petrographic and microtextural observations from blueschist facies pseudotachylytes, Cape Corse, Corsica. Tectonophysics 610, 138–149.

Doblas, M., 1998. Slickenside kinematic indicators. Tectonophysics 295, 187–197.

El-Wahed, M.A.A., Kamha, S.Z., 2010. Pan-African dextral transpressive duplex and flower structure in the Central Eastern Desert of Egypt. Gondwana Research 18, 315–336.

Federico, L., Crispini, L., Scambelluri, M., Capponi, G., 2007. Different PT paths recorded in a tectonic mélange (Voltri Massif, NW Italy): implications for the exhumation of HP rocks. Geodinamica Acta 20, 3–19.

Fodor, L., Sztanó, O., Kövér, Sz., 2013. Pre-conference field trip: mesozoic deformation of the northern transdanubian range (Gerecse and Vértes Hills). Acta Mineralogica-Petrographica. Field Guide Series 31, 1–34.

Fogarasi, A., 1995. Sedimentation on tectonically controlled submarine slopes of Cretaceous age, Gerecse Mts., Hungary - working hypothesis. Általános Földtani Szemle 27, 15–41.

Greiling, R.O., Garfunkel, Z., Zachrisson, E., 1998. The orogenic wedge in the central Scandinavian Caledonides: Scandian structural evolution and possible influence on the foreland basin. GFF 120, 181–190.

Hancock, P.L., 1985. Brittle microtectonics: principles and practice. Journal of Structural Geology 7, 437–457.

Hirth, G., Tullis, J., 1994. The brittle-plastic transition in experimentally deformed quartz aggregates. Journal of Geophysical Research: Solid Earth 99, 11731–11747.

Hoogerduijn Strating, E.H., Vissers, R.L.M., 1994. Structures in natural serpentinite gouges. Journal of Structural Geology 16, 1205–1215.

Jaroszewski, W., 1984. Fault and Fold Tectonics (Translated). J. Wiley & Sons, 565 p.

Karanth, R.V., Gadhavi, M.S., 2007. Structural intricacies: emergent thrusts and blind thrusts of central Kachchh, western India. Current Science 93, 1271–1280.

Karkanas, P., 1995. The slip-fiber chrysotile asbestos deposit in the Zidani area, northern Greece. Ore Geology Reviews 10, 19–29.

Katayama, I., Parkinson, C.D., Okamoto, K., Nakajima, Y., Maruyama, S., 2000. Supersilicicclinopyroxene and silica exsolution in UHPM eclogite and pelitic gneiss from the Kokchetav massif, Kazakhstan. American Mineralogist 85, 1368–1374.

Ludman, A., West, D.P., Jr., (Eds.), 1999. The Norumbega Fault System of the Northern Appalachians: Geological Society of America Special Paper, 331, 199 p.

Misra, A.A., Bhattacharya, G., Mukherjee, S., Bose, N., 2014. Near N-S paleo extension in the western Deccan region in India: Does it link strike-slip tectonics with India-Seychelles rifting? International Journal of Earth Sciences 1645–1680.

Misra, A.A., Sinha, N., Mukherjee, S., 2015. Repeat ridge jumps and microcontinent separation: insights from NE Arabian Sea. Marine and Petroleum Geology 59, 406–428

Mooney, W., Beroza, G., Kind, R., 2007. Fault zones from top to bottom. In: Handy, M.R., Hirth, G., Hovius, N. (Eds.), Tectonic Faults - Agents of Change on a Dynamic Earth. Dahlem Workshop Report 95. The MIT Press, Cambridge, Mass., USA, pp. 2–46.

Mukherjee, S., 2010a. Structures at meso- and micro-scales in the Sutlej section of the Higher Himalayan Shear Zone in Himalaya. e-Terra 7, 1–27.

Mukherjee, S., 2010b. Microstructures of the Zanskar shear zone. Earth Science India 3, 9–27.

Mukherjee, S., 2012a. Tectonic implications and morphology of trapezoidal mica grains from the Sutlej section of the Higher Himalayan Shear Zone, Indian Himalaya. The Journal of Geology 120, 575–590.

Mukherjee, S., 2012b. A microduplex. International Journal of Earth Sciences 101, 503.

Mukherjee, S., 2013a. Deformation Microstructures in Rocks. Springer.

Mukherjee, S., 2013b. Higher Himalaya in the Bhagirathi section (NW Himalaya, India): its structures, backthrusts and extrusion mechanism by both channel flow and critical taper mechanism. International Journal of Earth Sciences 102, 1851–1870.

Mukherjee, S., 2013c. Channel flow extrusion model to constrain dynamic viscosity and Prandtl number of Higher Himalayan Shear Zone. International Journal of Earth Sciences 102, 1811–1835.

Mukherjee, S., 2014a. Atlas of Shear Zone Structures in Meso-scale. Springer.

Mukherjee, S., 2014b. Review of flanking structures in meso- and micro-scales. Geological Magazine 151, 957–974.

Mukherjee, S., A review on out-of-sequence deformation in the Himalaya. In: Mukherjee, S., Carosi, R., van der Beek, P., Mukherjee, B.K., Robinson, D. (Eds.), Tectonics of the Himalaya. Geological Society, London. Special Publication 412 (in press).

Mukherjee, S., Koyi, H.A., 2010a. Higher Himalayan Shear Zone, Sutlej section: structural geology and extrusion mechanism by combination of simple shear, pure shear and channel flow in shifting modes. International Journal of Earth Sciences 99, 1267–1303.

Mukherjee, S., Koyi, H.A., 2010b. Higher Himalayan Shear Zone: Zanskar Indian Himalaya-Microstructural studies and extrusion mechanism by a combination of simple shear and channel flow. International Journal of Earth Sciences 99, 1267–1303.

Navabpour, P., Angelier, J., Barrier, E., 2007. Cenozoic post-collisional brittle tectonic history and stress reorientation in the High Zagros Belt (Iran, Fars Province). Tectonophysics 432, 101–131.

Novakova, L., 2010. Detailed brittle tectonic analysis of the limestones in the quarries near Vápenná village. Acta Geodynamica et Geomaterialia 7, 1–8.

Novakova, L., 2014. Evolution of paleostress fields and brittle deformation in Hronov-Poříčí Fault Zone. Bohemian Massif. Studia Geophysica et Geodaetica 58, 269–288.

Passchier, C.W., Trouw, R.A.J., 2005. Microtectonics. Springer, Berlin. 371 pp.

Pease, V., Argent, J., 1999. The northern Sacramento mountains, SW United States, Part I: structural profile through a crustal extensional detachment system. In: MacNiocaill, C., Ryan, P. (Eds.). Continental Tectonics, vol. 164. Geological Society of London Special Publication, pp. 179–198.

Pease, V., Foster, D., Wooden, J., O'Sullivan, P., Argent, J., Fanning, C., 1999. The northern Sacramento mountains, SW United States, Part II: exhumation history and detachment faulting. In: MacNiocaill, C., Ryan, P. (Eds.). Continental Tectonics, vol. 164. Geological Society of London Special Publication, pp. 199–237.

Petit, J.P., 1987. Criteria for the sense of movement on fault surfaces in brittle rocks. Journal of Structural Geology 9, 597–608.

Placer, L., Vrabec, M., Celarc, B., 2010. The bases for understanding of the NW Dinarides and Istria Peninsula tectonics. Geologija 53, 55–86.

Polino, R., Dal Piaz, G.V., Gosso, G., 1990. Tectonic erosion at the Adria margin and accretionary processes for the Cretaceous orogeny of the Alps. Mémoires. Societé Géologique France 156, 345–367.

Poljak, M., Živčić, M., Zupančič, P., 2000. The seismotectonic characteristics of Slovenia. Pure and Applied Geophysics 157, 37–55.

Price, N.A., Johnson, S.E., Gerbi, C.C., West Jr., D.P., 2012. Identifying deformed pseudotachylyte and its influence on the strength and evolution of a crustal shear zone at the base of the seismogenic zone. Tectonophysics 518–521, 63–83. http://dx.doi.org/10.1016/j.tecto2011.11.011.

Ramsay, J.G., Huber, R., 1983. The techniques of modern structural geology. Strain Analysis, vol. 1, Academic Press, New York, NY, p. 307.

Rattenbury, M.S., Townsend, D.B., Johnston, M.R., 2006. Geology of the Kaikoura Area. Scale 1:250 000. GNS Science, Lower Hutt, New Zealand.

Ravna, E.J.K., Andersen, T.B., Jolivet, L., De Capitani, C., 2010. Cold subduction and the formation of lawsoniteeclogite – constraints from prograde evolution of eclogitized pillow lava from Corsica. Journal of Metamorphic Geology 28, 381–395.

Rykkelid, E., Fossen, H., 2002. Layer rotation around vertical fault overlap zones: observations form seismic data, field examples, and physical experiments. Marine and Petroleum Geology 19, 181–192.

Sasvári, Á., Csontos, L., Palotai, M., 2009. Structural geological observations in tölgyhát quarry (Gerecse mts., Hungary). Földtani Közlöny 139, 55–66.

Sibson, R.H., Toy, V.G., 2006. The habit of fault-generated pseudotachylyte: Presence vs. absence of friction-melt. In: Earthquakes: Radiated energy and the physics of faulting. Geophysical Monograph Series 170, 153–166.

Simpson, C., 1985. Deformation of granitic rocks across the brittle-ductile transition. Journal of Structural Geology 7, 503–511.

Shimamoto, T., Togo, T., 2012. Earthquakes in the lab. Science 338, 54–55.

Sibson, R.H., 1975. Generation of pseudotachylyte by ancient seismic faulting. Geophysical Journal International 43, 775–794.

Smyth, J.R., 1980. Cation vacancies and the crystal chemistry of breakdown reactions in kimberlitic omphacites. American Mineralogist 65, 1185–1191.

Špaček, P., Sýkorová, Z., Pazdírková, J., Švancara, J., Havíř, J., 2006. Present-day seismicity of the south-eastern Elbe Fault System (NE Bohemian Massif). Studia Geophysica et Geodaetica 50, 233–258.

Spray, J., 1992. A physical basis for the frictional melting of some rock-forming minerals. Tectonophysics 204, 201–221.

Srivastava, D.C., John, G., 1999. Deformation in the Himalayan Frontal Fault Zone: evidence from small-scale structures in Mohand-Khara area, NW Himalaya. In: Jain, A.K., Manickavasagam (Eds.), Geodynamics of the NW Himalaya. Gondwana Research Group Memoir, pp. 273–284.

Swanson, M., 1992. Fault structure, wear mechanisms and rupture processes in pseudotachylyte generation. Tectonophysics 204, 223–242.

Tari, G., 1994. Alpine Tectonics of the Pannonian Basin (Ph.D. thesis). Rice University, Texas, USA, 501 p.

Tavarnelli, E., 1999. Normal faults in thrust sheets: pre-orogenic extension, post-orogenic extension, or both? Journal of Structural Geology 21, 1011–1018.

Teisseyre, H., 1959. Eingie Bemerkungen über die Methodik der Mikrostrukturen in der tektonischen Forschung, Freiberger Forsch., Hft., c.57.

Tjia, H.D., 1964. Slickensides and fault movements. Geological Society of America Bulletin. 75, 683–686.

Togo, T., Shimamoto, T., 2012. Energy partition for grain crushing in quartz gouge during subseismic to seismic fault motion: an experimental study. Journal of Structural Geology 38, 139–155.

Törő, B., Pratt, B.R., Renaut, R.W., 2013. Paleoseismic indicators in the lacustrine Green river formation (Eocene, USA) – characteristics and implications. Geological Society of America Abstracts with Programs 45, 357.

Tullis, T.E., Bürgmann, R., Cocco, M., Hirth, G., King, G.C.P., Oncken, O., Otsuki, K., Rice, J.R., Rubin, A., Segall, P., Shapiro, S.A., Wibberley, C.A.J., 2007. Group report: Rheology of fault rocks and their surroundings. In: Handy, M.R., Hirth, G., Hovius, N. (Eds.), Tectonic Faults - Agents of Change on a Dynamic Earth. Dahlem Workshop Report 95. The MIT Press, Cambridge, Mass., USA, pp. 183–204.

Valenta, J., Gazdova, R., Kolinsky, P., 2011. Seismo-hydrological monitoring in the area of the Hronov-Poříčí fault zone, northern Czech Republic, Central Europe. American Geophysical Union. Fall Meeting 2011, S11B–S2211.

Vanossi, M., Cortesogno, L., Galbiati, B., Messiga, B., Piccardo, G., Vannucci, R., 1984. Geologia delle Alpi Liguri: dati, problemi, ipotesi. Memorie della Società Geologica Italiana 28, 5–75.

Vignaroli, G., Rossetti, F., Belardi, G., Billi, A., 2011. Linking rock fabric to fibrous mineralisation: a basic tool for the asbestos hazard. Natural Hazards and Earth Science Systems 11, 1267–1280.

Vignaroli, G., Rossetti, F., Rubatto, D., Theye, T., Lisker, F., Phillips, D., 2010. Pressure-Temperature-Deformation-time (P-T-d-t) exhumation history of the Voltri Massif HP-complex, Ligurian Alps, Italy. Tectonics 29, TC6009.

Woldřich, J.N., 1901. Earthquake in the north-eastern Bohemia on January 10, 1901 [in Czech]. Transaction of the Academic Sciences of the Czech Republic, Series II 10, 1–33.

World Health Organization, 1986. Asbestos and Other Natural Mineral Fibres. Environmental Health Criteria, Geneva. No. 53.

Yao, L., Shimamoto, T., Ma, S., Han, R., Mizoguchi, K., 2013. Rapid postseismic strength recovery of Pingxi fault gouge from the Longmenshan fault system: experiments and implications for the mechanisms of high-velocity weakening of faults. Journal of Geophysical Research 118, 1–17, http://dx.doi.org/10.1002/jgrb.50308.

Chapter 4

Boudins and Mullions

KEYWORDS
Boudins; Mullions; Pinch and swell structures; Scar folds.

Local brittle–ductile extension partially or completely separate clasts. These segmented clasts are called boudins. Asymmetric boudins can be used to decipher shear sense (Goscombe et al., 2004). See Abe et al. (2013) for fracture patterns in boudinage. Mullions are *linear fluted structures developed within a rock or at lithological interfaces* (Twiss and Moores, 2007). Depending on their geneses, Twiss and Moores (2007) classified them into three types: "fold mullion," "fault mullion," and "irregular mullion." Viscosity contrast between the boudinaged/mullion material and the host rock is one of the controlling factors of their geometries (Sokoutis, 1990; Talbot, 1999; Schmalholz et al., 2008; Schmalholz and Maeder, 2012). Maeder et al. (2009) used the term "segment structure" to describe boudins and mullions together (Figs. 4.1–4.18). Mukherjee (2014a,b) reviewed terms to describe boudins in the light of flanking structures.

FIGURE 4.1 **Photograph from the Gaub Canyon, Namibia, showing a boudinaged metabasalt layer enveloped by metaturbidites in the Southern Marginal Zone of the Damara Belt.** The rocks deformed during Neoproterozoic to Cambrian, Pan African assembly of Gondwana (e.g., Barnes and Sawyer, 1980), with peak metamorphic conditions in the amphibolite facies. The rock assemblage comprises metamorphosed basalt, sandstone, mudstone, and chert, which represent intercalated continental and oceanic materials that were likely deformed in an accretionary prism (Kukla and Stanistreet, 1991). The photo highlights the rheological contrasts that develop through intercalation of lithologies of variable viscosity. In this case, the basaltic layer has fractured in the boudin neck, allowing formation of a quartz vein, whereas metasediments are locally folded to fill the gap between boudins of metabasalt. This provides an example of mixed brittle–viscous deformation at temperatures in excess of the brittle to viscous transition in quartzofeldspathic rocks, and a possible geological analogue to the mixture of transient creep and episodic brittle failure inferred to be responsible for deep tremor and slow slip in active subduction zones (Fagereng et al., 2014). Photo width: ~2 m. *(Ake Fagereng)*

Atlas of Structural Geology. http://dx.doi.org/10.1016/B978-0-12-420152-1.00004-1

FIGURE 4.2 **Boudinaged quartz vein within meta-greywacke.** Kumbhalgarh Formation, ~85 km northwest of Udaipur, Rajasthan, India. *(Swagato Dasgupta)*

FIGURE 4.3 **A set of composite boudins (Ghosh, 1993; Ghosh and Sengupta, 1999) with two orders of boudinage.** The rock is a migmatite with interlayered amphibolites and quartzofeldspathic gneisses. The small boudins are nested within the larger boudins of the multilayered unit. A layer-normal compression stretched and boudinaged the more competent thinner layers of amphibolites in the central part. With continued deformation, the multilayer as a whole extended and successively boudinaged into composite boudins. Oppositely oriented bending folds near the separation zone of the larger boudins. Pen length: 15 cm. Chhotonagpur Gneissic Complex, Ratanpahar, Jasidih, Jharkhand, India. *(Sudipta Sengupta, Sadhana M. Chatterjee)*

FIGURE 4.4 **Boudinaged amphibolite layer within a migmatite gneiss sequence on the shoreline of northern Scotland, near Durness.** Dilatation space is filled by two different rock types: migmatite gneiss layers are symmetrically inflected toward the core. Granite melt from the surrounding migmatite is attracted by the boudin neck free space. *(Guido Gosso)*

FIGURE 4.5 **Pinch and swell structures ~3 ft long.** The swells are subelliptical. Rock types: sandstone and mudstone. From western side of Rakhavdav complex, Udaipur district, Rajasthan, India. *(Moloy Sarkar)*

FIGURE 4.6 **Pinch and swell structures in two metabasic dykes discordantly hosted by high-grade gneisses on the shoreline facing the Italian Antarctic base at Terra Nova Bay, Northern Victoria Land, Antarctica.** Gentle ductile shortening of white veins transecting the thicker dyke neck zone indicate veining penecontemporaneous to pinching and synchronous brittle and ductile deformation in the dyke. A geologist in yellow pants stands for scale atop the scree. *(Guido Gosso)*

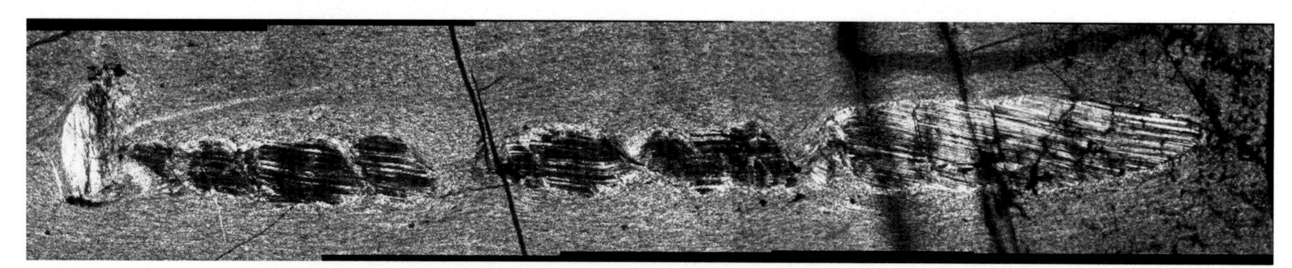

FIGURE 4.7 **Microboudinage structure in an ultramylonitized peridotite.** The boudinage occurred in a mantle orthopyroxene deformed on the [001] (100) slip system (e.g., Christensen and Lundquist, 1982). The boudin fragments show patched undulose extinction, subgrains, and are mantled by fine recrystallized grains. Note that the boudin fragment, at left, is rotated 90° with respect to the overall orientation of the clasts. This might make the flow heterogeneous at the edge of the clast. Top-to-left shear. Location: Archipelago of Saint Peter and Saint Paul (Brazil). Crossed polarized light. Width of the view: 7 mm. *(Suellen Olívia Cândida Pinto, Leonardo Evangelista Lagoeiro, Luiz Sérgio Amarante Simões, Paola Ferreira Barbosa)*

FIGURE 4.8 **Chocolate tablet boudins, Lameta Ghat, Jabalpur, Madhya Pradesh, India.** *(Sreejita Chatterjee)*

FIGURE 4.9 **A quartz vein boudinaged into nearly rectangular units.** Jabalpur, Madhya Pradesh, India. *(Sreejita Chatterjee)*

FIGURE 4.10 **Rectangular boudinaged quartz veins in calc-schist.** Sense of slip of the boudinaged clast near the hammer is top-to-right. South Delhi Fold Belt. Taleti village, Ambaji, Gujarat, India. See Mukherjee (2010a,b, 2013a,b, 2014a,b) and Mukherjee and Koyi (2010a,b) for other boudin varieties from different shear zones. *(Narayan Bose, Soumyajit Mukherjee)*

FIGURE 4.11 **Boudinaged quartz veins in calc-schist.** One of the boudinaged quartz clast near the center of the photograph is deformed internally and deviates significantly from rectangular shape. South Delhi Fold Belt. Taleti village, Ambaji, Gujarat, India. *(Narayan Bose, Soumyajit Mukherjee)*

FIGURE 4.12 **Asymmetric rotating domino boudins within low-grade metamorphic carbonate sequence.** The domino boudins are made up of dolomite, while the surrounding rock is calcite marble. At low-grade conditions, dolomite intercalations act as more competent layers. The calcite marble deforms crystal plastically forming foliated calcite mylonite. The intercalated dolomite layers remain nearly rigid and disrupt into rectangular boudins. Rectangular shape of the dolomite dominos indicates high viscosity contrast during deformation (Fossen, 2010). The domino boudins are asymmetric ones and parallel mylonitic foliation. The individual dominos are separated by small-scale shear zones that die out as soon as they leave the dolomite layer. Some dilatation occur across the inter-boudin surfaces. Significant rotation of the domino boudins are present, which kept the row of boudins aligned with the general foliation of the calcite mylonite. The stretch magnitude varies along the boudin layers, but usually remains low. Rarely domino boudins separate completely. Asymmetric domino boudins indicate sinistral shear. Other shear criteria in field confirmed this. However, asymmetric boudins can be problematic shear-sense indicators (Goscombe and Passchier, 2003), thus require additional studies. Triassic limestone and dolomite (Schefer et al., 2010). Photo width ~20 cm. Brzeće, Kopaonik Mts., SE Serbia. Coordinates: N43°18′ 44.05″, E20°50′59.75″. *(László Fodor)*

FIGURE 4.13 **Mullions and vertical columns with their characteristic highly penetrative nature in Neoproterozoic quartzites exposed in the western border of the Kalahari Desert, southeast region of Angola, near Vipongo area.** The elongate columns forming the mullions are bound by planes of subvertical northeast–southwest cleavages probably associated with regional transposed folds. Their cylindrical geometry associate with intense stretching of the rock parallel to the mullions by constriction. These structures in the Kalahari Desert call attention with their appearance of stack of telegraph poles or even a cut forest with their shopped vertical tree trunks spread along the hill side. Strong brittle to brittle–ductile cleavages in quartzites developed by partitioned sinistral transpression. See Figs. 4.15–4.17 for more examples. *(Roberto Vizeu Lima Pinheiro)*

FIGURE 4.14

FIGURE 4.15

FIGURE 4.16

FIGURE 4.17 **Mullions within Precambrian orthoquartzite of Aravalli Supergroup.** Zawar mine area, Udaipur, Rajasthan, India. *(Rohini Das)*

FIGURE 4.18 **Gray conglomerate overlying red mud horizon at Wine Strand, Ballyferriter, Co. Kerry, Ireland.** A flame-type structure is present at the boundary between the units. Is it of primary sedimentary origin due to a density gradient? Or is it a mullion structure of tectonic origin? *(Kieran F. Mulchrone, Patrick Meere)*

REFERENCES

Abe, S., Ural, J., Kettermann, M., 2013. Fracture patterns in nonplane strain boudinage—insights from 3-D discrete element models. Journal of Geophysical Research: Solid Earth 118, 1304–1315.

Barnes, S., Sawyer, E., 1980. An alternative model for the Damara mobile belt: Ocean crust subduction and continental convergence. Precambrian Research 13, 297–336.

Christensen, Nikolas I., Lundquist, Susan M., 1982. Pyroxene orientation within the upper mantle. The Geological Journal of America Bulletin 93, 282.

Fagereng, A., Hillary, G.W.B., Diener, J.F.A., 2014. Brittle-viscous deformation, slow slip, and tremor. Geophysical Research Letters 41, 4159–4167.

Fossen, H., 2010. Structural Geology. Cambridge University Press, New York.

Ghosh, S.K., 1993. Structural Geology: Fundamentals and Modern Developments. Pergamon Press, Oxford. 598 pp.

Ghosh, S.K., Sengupta, S., 1999. Boudinage and composite boudinage in superposed deformation and migmatization. Journal of Structural Geology 21, 97–110.

Goscombe, B.D., Passchier, C.W., 2003. Asymmetric boudins as shear sense indicators—an assessment from field data. Journal of Structural Geology 25, 575–589.

Goscombe, B.D., Passchier, C.W., Hand, M., 2004. Boudinage classification, end member boudin types and modified boudin structures. Journal of Structural Geology 26, 739–763.

Kukla, P.A., Stanistreet, I.G., 1991. Record of the Damaran Khomas Hochland accretionary prism in central Namibia: refutation of an "ensialic" origin of a Late Proterozoic orogenic belt. Geology 19, 473–476.

Maeder, X., Passchier, C.W., Koehn, D., 2009. Modelling of segment structures: boudins, bone-boudins, mullions and related single- and multiphase deformation features. Journal of Structural Geology 31, 817–830.

Mukherjee, S., 2010a. Microstructures of the Zanskar Shear Zone. Earth Science India 3, 9–27.

Mukherjee, S., 2010b. Structures at meso- and micro-scales in the Sutlej section of the Higher Himalayan Shear Zone in Himalaya. e-Terra 7, 1–27.

Mukherjee, S., 2013a. Deformation Microstructures in Rocks. Springer.

Mukherjee, S., 2013b. Higher Himalaya in the Bhagirathi section (NW Himalaya, India): its structures, backthrusts and extrusion mechanism by both channel flow and critical taper mechanism. International Journal of Earth Sciences 102, 1851–1870.

Mukherjee, S., 2014a. Review of flanking structures in meso- and micro-scales. Geological Magazine 151, 957–974.

Mukherjee, S., 2014b. Atlas of Shear-zone Structures in Meso-scale. Springer.

Mukherjee, S., Koyi, H.A., 2010a. Higher Himalayan Shear Zone, Zanskar Indian Himalaya-microstructural studies and extrusion mechanism by combination of simple shear and channel flow. International Journal of Earth Sciences 99, 1083–1110.

Mukherjee, Koyi, 2010b. Higher Himalayan Shear Zone, Sutlej section: structural geology and extrusion mechanism by combination of simple shear, pure shear and channel flow in shifting modes. International Journal of Earth Sciences 99, 1267–1303.

Schefer, S., Egli, D., Missoni, S., Bernoulli, D., Fügenschuh, B., Gawlick, H.-J., Jovanović, D., Krystyn, L., Lein, R., Schmid, S.M., Sudar, M., 2010. Triassic metasediments in the internal Dinarides (Kopaonik area, southern Serbia): stratigraphy, paleogeographic and tectonic significance. Geologica Carpathica 61, 89–109.

Schmalholz, S.M., Maeder, X., 2012. Pinch-and-swell structure and shear zones in viscoplastic layers. Journal of Structural Geology 37, 75–88.

Schmalholz, S.M., Schmid, D.W., Fletcher, R.C., 2008. Evolution of pinch-and-swell structures in a power-law layer. Journal of Structural Geology 30, 649–663.

Sokoutis, D., 1990. Experimental mullions at single and double interfaces. Journal of Structural Geology 12, 365–373.

Talbot, C.J., 1999. Can field data constrain rock viscosities? Journal of Structural Geology 21, 949–957.

Twiss, R.J., Moores, E.M., 2007. Structural Geology, second ed. W. H. Freeman and Company, New York. 313 pp.

Chapter 5

Veins

KEYWORDS
Cross-cutting veins; Sigmoid veins; Veins.

Crosscutting veins found in almost all rock types are perhaps the easiest structures to pick up by novices. One can deduce stress directions by studying attitude of veins (Fossen, 2010). Veins develop commonly orthogonal to the maximum instantaneous extension axis, and have been considered as a good shear sense indicator (Davis and Reynolds, 1996). Thus, careful study of complex vein systems has been proved fruitful in tectonic studies (Maeder et al., 2014). Passchier and Trouw (2005) described veins morphologically as fibrous and massive. These two types of veins can have either sharp or fuzzy margins. Leucosome veins cutting across alternate leucosome and melanosome layers are common in migmatites (Mukherjee, 2010, 2014). A recent review on veins of tectonic origin is available in Bons et al. (2012) (Figures 5.1–5.8).

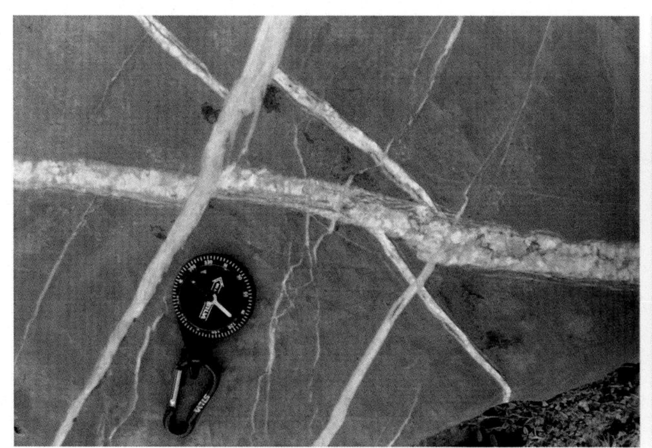

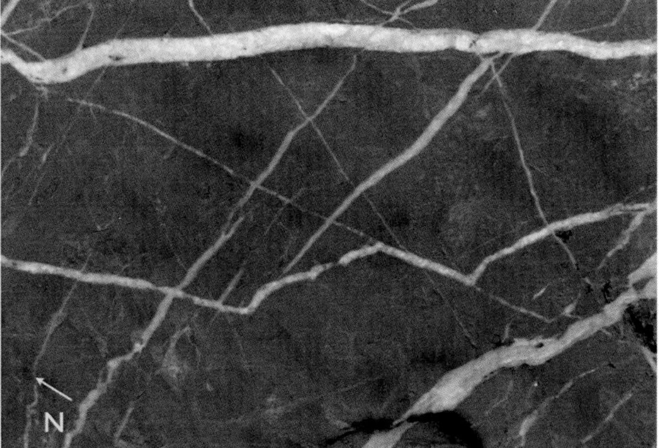

FIGURE 5.1 **Vein interaction in a high-pressure cell in limestone.** Veins are common in carbonates on the southern flank of Jabal Akhdar Anticline in the Oman Mountains. The area exposes the 'Autochthonous' Hajar Supergroup, deposited in a predominantly passive margin environment and subject to a multiphase deformation related to the convergence of the Arabian with the Eurasian plate, when the Hawasina and the Semail Ophiolite nappes were emplaced onto the Arabian plate margin. This deformation resulted in several successive generations of calcite veins, stylolites and calcite cemented faults formed in a changing stress regime, in a near- lithostatic fluid pressure regime at several km depth, at temperatures over 200 °C. These are exposed in spectacular fashion and quality in a very large area, being perhaps the best exposed complex vein system on Earth. These two outcrop photos illustrate veins at high angle to bedding, in the Natih formation close to the contact with the overlying Muti formation. The first photo shows reasonably clear crosscutting relationships between three generations of veins at high angle to bedding. Photo was taken looking down on the bedding surface. Veins of different generations have different morphologies, different amounts of solid inclusions, and different degree of dissolution by later stylolites. We leave the reader to determine the relative sequence of the structures. The second picture shows an example of different vein generations, where overprinting relationships are not so clear. We interpret these patterns to be caused by the interaction of propagating fractures with preexisting veins, causing the fractures to reorient, following mechanical heterogeneities. See Hilgers et al. (2006), Holland et al. (2009a,b), Holland and Urai (2010), and Virgo et al. (2013) for detail. First picture: 23°14′48.23″N 57°9′15.67″E Width of image is about 30 cm. Second picture: 23°13′58.85″N 57°9′47.05″E Width of image is about 30 cm. *(Janos L. Urai, Max Arndt, Simon Virgo)*

Atlas of Structural Geology. http://dx.doi.org/10.1016/B978-0-12-420152-1.00005-3

FIGURE 5.2 **Veins related to dilatant faults are common in carbonates on the southern flank of Jabal Akhdar Anticline in the Oman Mountains.** The area exposes the 'Autochthonous' Hajar Supergroup, deposited in a predominantly passive margin environment and subject to a multiphase deformation related to the convergence of the Arabian with the Eurasian plate, when the Hawasina and the Semail Ophiolite nappes were emplaced onto the Arabian plate margin. This deformation resulted in several successive generations of calcite veins, stylolites and calcite cemented faults formed in a changing stress regime, in a near- lithostatic fluid pressure regime at several km depth, at temperatures over 200 °C. These are exposed in spectacular fashion and quality in a very large area, being perhaps the best exposed complex fault system on Earth. The pictures show profile views (looking along bedding, to the west) of an incipient normal fault in the damage zone of a larger fault which has an offset of several tens of m, with striations indicating normal, oblique-slip and strike slip motion. Location is in Wadi Nakhr, the Grand Canyon of Oman. The fault in the outcrop clearly displaces an earlier set of bedding-parallel veins, and also a set of en-echelon veins which probably grew in the early stages of the deformation phase leading to the fault. The outcrop shows segmentation, drag of bedding, and a fault branch. On the right side of the picture is a weathered - out normal fault, defining a small graben structure. Location Fault Wadi Nakhr 23°11′2.15″N 57°12′23.81″E Width of first picture is about 4 m. Second picture is detail of the first one. For detail, see Holland et al. (2009a,b). *(Janos L. Urai, Simon Virgo, Max Arndt)*

FIGURE 5.3 **Semi-ductile shear zone in limestone.** Early Permian dark nodular limestones with intervening thin layers of shale were first affected by late-diagenetic pressure solution along the bedding planes (Novak, 2007). A small-scale dextral shear zone offsets subvertical limestone beds for a few decimeters. Shearing was absorbed by opening of en-echelon extensional veins oblique to the shear zone along principal shortening, and by pressure solution perpendicular to principal shortening direction, evidenced by markedly stylolitic solution planes, progressive thinning of beds toward the center of the shear zone, and by abrupt truncations of extensional veins along solution seams. This is a nice example of semi-ductile deformation of carbonates under nonmetamorphic conditions, and presents a textbook example of geometrical arrangement of minor structures in shear zones (cf. Davis et al., 2012). Rock type: thin-bedded nodular limestone with shale intercalations, Born Formation, early Permian. Field of view: ~1 m. Dovžan gorge, Karavanke mountains, Slovenia. Coordinates: 46°19′47″ N, 14°19′47″ E. *(Marko Vrabec)*

FIGURE 5.4 **A quartz vein cut across gneissic foliations.** Note some of the quartzofeldspathic layers are connected with the vein. The gneissose foliation is warped, and we cannot confirm whether the foliation is curved due to any rotation of the vein. Higher Himalaya, Jhakri, Rampur district, Himachal Pradesh, India. *(Soumyajit Mukherjee)*

FIGURE 5.5 **A quartz-rich vein in a mylonitized gneiss. The quartzofeldspathic foliation within the host rock is connected with the vein.** Except the bottom part of the vein, no clear-cut ductile shear sense can be deciphered from this exposure. Higher Himalaya, Jhakri, Rampur district, Himachal Pradesh, India. *(Soumyajit Mukherjee)*

FIGURE 5.6 **The bottom part of the previous figure is zoomed.** Note a top-to-left (down) ductile shear sense deciphered near the bottom part of the sheared vein. Higher Himalaya, Jhakri, Rampur district, Himachal Pradesh, India. See Mukherjee (2007, 2010, 2013a,b) and Mukherjee and Koyi (2010) for regional geology. See Mukherjee (2014) for more such structures. *(Soumyajit Mukherjee)*

FIGURE 5.7 **A thin-curved quartz vein within sheared gneiss, possibly top-to-right sheared.** Foliation of the host rock is warped and does not show clear-cut ductile shear sense. Higher Himalaya, Jhakri, Rampur district, Himachal Pradesh, India. *(Soumyajit Mukherjee)*

FIGURE 5.8 **Quartz vein network defined mainly by two sets of veins.** The set of veins dipping toward left are usually thicker than that dipping toward right. Chitradurga district, Karnataka, India. *(Aabha Singh)*

REFERENCES

Bons, P.D., Elburg, M.A., Gomez-Rivas, E., 2012. A review of the formation of tectonic veins and their microstructures. Journal of Structural Geology 43, 33–62.

Davis, G.H., Reynolds, S.J., 1996. Structural Geology of Rocks and Regions, second ed. John Wiley & Sons, Inc.

Davis, G.H., Reynolds, S.J., Kluth, C.F., 2012. Structural Geology of Rocks and Regions, third ed. John Wiley & Sons. 839p.

Fossen, H.A., 2010. Structural Geology. Cambridge University Press. pp. 122–123.

Hilgers, C., Kirschner, D., Breton, J.-P., Urai, J.L., 2006. Fracture sealing and fluid overpressures in limestones of the Jabel Akhbar Dome, Oman Mountains. Geofluids 6, 168–184.

Holland, M., Urai, J.L., 2010. Evolution of anastomosing crack–seal vein networks in limestones: insight from an exhumed high-pressure cell, Jabal Shams, Oman Mountains. Journal of Structural Geology 32, 1279–1290.

Holland, M., Urai, J.L., Muchez, P., Willemse, J.M., 2009a. Evolution of fractures in a highly dynamic thermal, hydraulic, and mechanical system; (I), Field observations in Mesozoic carbonates, Jabal Shams, Oman Mountains. GeoArabia 14, 57–110.

Holland, M., Saxena, N., Urai, J.L., 2009b. Evolution of fractures in a highly dynamic thermal, hydraulic, and mechanical system—(ii) remote sensing fracture analysis, Jabal Shams, Oman Mountains. GeoArabia 14, 163–194.

Maeder, 2014. Complex vein systems as a data source in tectonics: An example from the Ugab valley, NW Namibia. Journal of Structural Geology 62, 125–140.

Mukherjee, S., 2007. Geodynamics, Deformation and Mathematical Analysis of Metamorphic Belts of the NW Himalaya. (Unpublished Ph.D. thesis). Indian Institute of Technology Roorkee, pp.1–267.

Mukherjee, S., 2010. Structures in meso- and micro-scales in the Sutlej section of the Higher Himalayan Shear Zone, Indian Himalaya. e-Terra 7, 1–27.

Mukherjee, S., 2010a. Structures at Meso- and Micro-scales in the Sutlej section of the Higher Himalayan Shear Zone in Himalaya. e-Terra 7, 1–27.

Mukherjee, S., 2013a. Higher Himalaya in the Bhagirathi section (NW Himalaya, India): its structures, backthrusts and extrusion mechanism by both channel flow and critical taper mechanism. International Journal of Earth Sciences 102, 1851–1870.

Mukherjee, S., 2013b. Channel flow extrusion model to constrain dynamic viscosity and Prandtl number of Higher Himalayan Shear Zone. International Journal of Earth Sciences 102, 1811–1835.

Mukherjee, S., 2014. Atlas of Shear Zone Structures in Meso-scale. Springer.

Mukherjee, S., Koyi, H.A., 2010a. Higher Himalayan Shear Zone, Sutlej section: structural geology and extrusion mechanism by various combination of simple shear, pure shear and channel flow in shifting modes. International Journal of Earth Sciences 99, 1267–1303.

Novak, M., 2007. Depositional environment of upper carboniferous—lower permian beds in the Karavanke mountains (Southern Alps, Slovenia). Geologija 50, 247–268.

Passchier, C.W., Trouw, R.A.J., 2005. Microtectonics, second ed. Springer.

Virgo, S., Abe, S., Urai, J.L., 2013. Extension fracture propagation in rocks with veins: insight into the crack-seal process using Discrete Element Method modeling. Journal of Geophysical Research: Solid Earth 118, 5236–5251.

Chapter 6

Various Structures

KEYWORDS
Columnar joints; Crater; Fissures; Fractures; Pull apart structure; Spheroidal weathering; Xenoliths.

This chapter presents various structures some of which are not worked intensely by structural geologists. Grain boundary migration in rocks observed under optical microscope can constrain the temperature the rock underwent (Stipp et al., 2002). Stability of buildings and fracturing during earthquakes has been a research topic to geoscientists (Krishnan et al., 2006). Study of faults and other structures has helped geoscientists to paleohydrology (Treiman, 2008) (Figures 6.1–6.60).

FIGURE 6.1 **Columnar joints, colonnade and entablature.** Thick solidifying lava flows develop contraction fractures that propagate from their cooling margins toward hotter interiors. These fractures, called columnar joints, divide a lava flow into columns, with polygonal (ideally hexagonal) shapes in plan. A subhorizontal basaltic lava flow of the Talisker Bay Group, Isle of Skye, Scotland is shown in the figure. The lower part of the flow shows well-developed vertical columns, suggesting nearly horizontal isotherms (contours of constant temperature within the lava flow). Such a columnar tier is called a colonnade. The upper part of this flow shows a highly chaotic and distorted internal structure, as would develop during rainfall or stream flow supplying water into the cooling flow interior and disturbing isotherms. This tier is known as an entablature. The colonnade and the entablature, though with greatly different field appearance, are part of a single lava flow. Here, the combined thickness of both is 120m. *(Hetu Sheth)*

Atlas of Structural Geology. http://dx.doi.org/10.1016/B978-0-12-420152-1.00006-5

FIGURE 6.2 **Distinct colonnade and entablature tiers of a single basaltic lava flow from the Kizilcahamam volcanics near Ankara, Turkey.** The flow dips toward left. Individual lava flows can have multiple tiers of this kind, depending on the exact mode of emplacement and environmental conditions. *(Hetu Sheth)*

FIGURE 6.3 **Curved columnar joints (entablature) in Bagaces Formation ignimbrites, Guanacaste province, Costa Rica.** The columns are curved and have six parallel sides (hexagonal) with heights between 1 and 5 m, and width between 15 and 20cm. The columns are 45m high. See Denyer and Alvarado (2007) for detail. 10°39′25.52,244″N, 85°37′59.70,504″W. *(Guido Sibaja Rodas)*

FIGURE 6.4 **Southwestward panoramic view of the geothermal travertine area of Bridgeport, California.** Photographed from the crest of the Hot Tub Ridge, which is an active and prominent fissure-ridge travertine deposit (Latitude N38°14′45″, Longitude W119°12′18″; Chesterman and Kleinhampl, 1991). The Hot Tub Ridge is 84 m long, 7 m wide, and 4 m high (De Filippis and Billi, 2012). The Long Ridge (a 360 m long fissure-ridge) and the Sierra Nevada mountains are also visible in the photograph. Fissure-ridge travertines are elongated mound-shaped deposits of travertine developed along open fissures usually in active geothermal-tectonic areas. The ridges can be straight, slightly curved or bifurcated in plan view, and are usually characterized by an axial extensional fissure extending along the crest of the ridge. The fissure-ridge in the photograph is unique for its singular tripartition of the fissure at the ridge tip. In active fissure-ridges, carbonate-rich hot waters ascending along the axial fissure cause carbonate precipitation both within the fissure-ridge and over the ridge flanks, thus generating banded and bedded travertine deposits, respectively (Hancock et al., 1999). Most fissure-ridges are located on the hanging wall of normal faults. Influence of active tectonics on the growth of fissure-ridge travertines may be moderate to important. Physicochemical attributes of fluids as well as their abundance (pore pressure) and climate oscillations (dry vs wet periods) also play an important role in the fissure-ridge nucleation and growth (Chesterman and Kleinhampl, 1991; Hancock et al., 1999; De Filippis and Billi, 2012). *(Luigi De Filippis, Andrea Billi)*

FIGURE 6.5 **Southeastward panoramic view of the Kamara active fissure-ridge, in the Denizli extensional basin, southwest Turkey (Latitude N38°03′24′, Longitude E28°58′16′; De Filippis et al., 2012, 2013).** An open axial fissure is well exposed along the ridge crest and is partly filled by banded travertine dated through U–Th methods (1.7±0.1 and 2.5±0.1 ka; De Filippis et al., 2012). The ridge crest is also characterized by a system of en-echelon open subvertical fractures with an average strike ~N120°. Fracture aperture varies between approximately a few millimeters and a maximum of c. 20 cm, and it is usually larger in the central section of the fissure-ridge than near its lateral closures. This fissure-ridge is 63 m long, 15 m wide, and 6 m high. Dimensions of known fissure-ridges on the Earth vary (e.g., Brogi and Capezzuoli, 2009) up to a maximum of ~2 km long, 400 m wide, and 20 m high (De Filippis et al., 2012). Both flanks of the Kamara fissure-ridge consist of bedded travertines dipping away from the axial fissure. This fissure-ridge is singular for the presence of a fossil waterfall along northeast flank and for the marked transverse asymmetry of the ridge. *(Luigi De Filippis, Andrea Billi)*

FIGURE 6.6 **This photograph shows an exposure across the southeast flank of the Long Ridge (Latitude N38°14′43″, Longitude W119°12′19″), which is a fissure-ridge located in the geothermal area of Bridgeport (California).** A subvertical banded travertine (dotted yellow lines) intruded within the axial sector of the fissure-ridge, whose flanks are mostly formed by steep travertine beds (dotted turquoise lines). The bedded travertine is a porous and stratified flowstone deposit forming the bulk (flanks) of fissure-ridges. The banded travertine, in contrast, is a nonporous, sparitic travertine filling the interior of fissure ridges with steep-to-vertical thick veins in the axial region and sill-like structures along the bedded travertine. The subvertical banded travertine usually cuts across the preexisting bedded travertine, which, in turn, may upward suture the banded travertine, thus providing evidence for the alternate growth of banded and bedded travertines. The ridge is unique because it exposes this reciprocal crosscutting and suturing relationships between banded and bedded travertines. From a paleoclimatic point of view, the bedded travertine growth is presumably connected with high stands of the water table during warm-humid periods (Faccenna et al., 2008), whereas the banded travertine growth is probably driven by coseismic exsolution events during low stands of the water table in cold-dry conditions (Uysal et al., 2009). Fissure-ridges can thus provide unique information on the relationship between paleoclimate, groundwater systems, and tectonics (Faccenna et al., 2008; Uysal et al., 2009; Crossey and Karlstrom, 2012). *(Luigi De Filippis, Andrea Billi)*

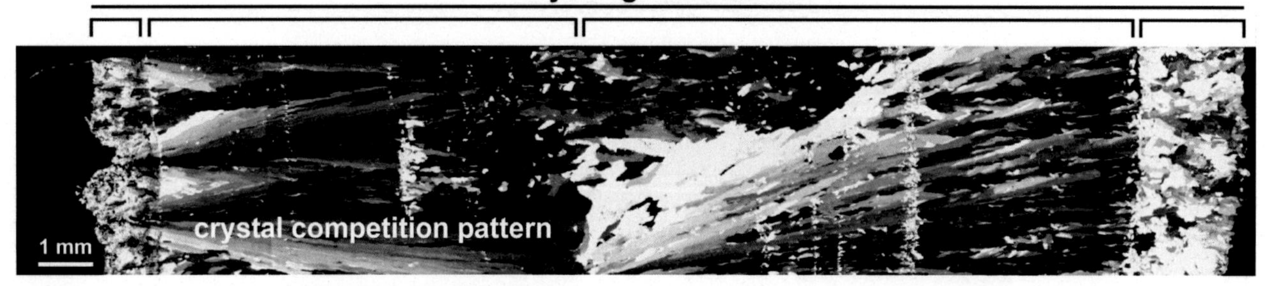

FIGURE 6.7 **Microscopic view of a vein of banded travertine across the Akköy fissure-ridge (Denizli basin, Turkey, Latitude N 37°56′56″, Longitude E 29°05′28″).** Main features are a series of long fibrous calcite (and subordinate aragonite) crystals with a typical competition pattern (feather-like structures). Multiple succeeding bands of crystals attest for the multiphase growth history of the veins (Uysal et al., 2007; Van Noten et al., 2013). The crystals are usually perpendicular to the vein walls and their growth typically occurred toward the center of the vein (symmetric syntaxial growth) or from one wall toward the other wall (asymmetric syntaxial growth), as also confirmed by radiometric dating (Uysal et al., 2007). The force of crystallization along the veins can induce the uplift of the rock above the veins, in the case of sill-like veins (Gratier et al., 2012), and the transverse opening of fissure-ridges, in the case of subvertical veins developed along the axial region of the ridges. *(Luigi De Filippis, Andrea Billi)*

FIGURE 6.8 **Extensional fractures with oriented calcite infill adjacent to fault core in limestone.** Alpine foreland basin, Clumanc, France. *(Amy Ellis)*

FIGURE 6.9 **Fold-and-thrust system.** Photo taken inside the Ordesa and Monte Perdido National Park (Spanish Pyrenees) to the North, in Llanos de Salarons. View of the Pico Blanco (2919 m). 30T 740,355.09 E 4,728,955.97 N. Paleocene and Eocene limestones (Salarons and Gallinera Formations) in gray (Ternet et al., 2008). Four sheets of the Monte Perdido fold-and-thrust system consisting of anticlines associated with south-verging thrusts. Mechanically weak layers inside the Upper Cretaceous rocks act as detachment levels for these thrusts (Séguret, 1972). The absence in this sector of the Southern Pyrenees of the Upper Triassic Keuper shales, evaporites, and salts (i.e., the regional detachment level) could control the genesis and geometry of the Monte Perdido fold-and-thrust system (Muñoz et al., 2013). *(Ruth Soto)*

FIGURE 6.10 Fracture planes in sandstones. Photo taken inside the Ordesa and Monte Perdido National Park (Spanish Pyrenees), in Carriata. 30T 740703.38 E 4727360.52 N. Campanian-Maastrichtian sandstones (Marboré Formation) (Ternet et al., 2008) showing spectacular vertical fracture planes oriented NNW-SSE and perpendicular to horizontal bedding. In the Pyrenees, shortening was accommodated by the formation of basement and cover thrust systems, and also by cleavage development of Alpine age (Choukroune and Séguret, 1973). One of the effects of the Pyrenean compression is the formation of this kind of fractures, that greatly influences on the Ordesa and Monte Perdido National Park landscape together with glacial and karst processes. *(Ruth Soto, Esther Izquierdo-Llavall)*

FIGURE 6.11 Thrust sheet stacking. Photo at Spanish French border to the south. View of the Ordesa and Monte Perdido National Park (Spanish Pyrenees), Monte Perdido peak (3355 m), Monte Perdido glacier, and Marboré lake. 31T 257569.12 E 4731539.75 N. Campanian–Maastrichtian sandstones (Marboré Formation) in brown-gray and Paleocene and Eocene limestones (Salarons and Gallinera Formations) in gray (Ternet et al., 2008). The Monte Perdido thrust sheet stacking, at the northern end of the South Pyrenean fold-and-thrust belt, consists of a beautiful and spectacular example of thin-skinned fold-and-thrust system that forms the highest calcareous massif in Western Europe. It is located at the southern end of the Pyrenean Axial Zone and represents, due to its spectacular outcrops, a key area to understand the Pyrenean mountain building. With a WNW-ESE trend, it has 6 km of estimated southwards horizontal displacement affecting cover sediments (Séguret, 1972). *(Ruth Soto)*

FIGURE 6.12 **Shear-enhanced compaction bands.** Bhuj Sandstone, Jawaharnagar,Kachchh, Gujarat, India. They formed at shallow burial conditions during the Kachchh basincompressional phase. These bands are seen on the backlimb zone of faultrelated fold, along E-W striking thrust faults. The compaction bands aredisplayed as thrust splays (arrow). One of the compaction band on theextreme right side display "S" shaped feature. Scale person (circled) height: 165 cm. *(George Mathew)*

FIGURE 6.13 **Photograph from the Matchless amphibolite, Kuiseb Canyon, Namibia, showing a lens ("phacoid") of metabasalt enveloped by metapelite.** The Matchless amphibolite is a hundreds of meters thick zone of predominantly mafic rocks, semicontinuous for hundreds of kilometers along strike, at constant structural level within the Southern Zone of the Pan African Damara belt. The rocks experienced peak metamorphic conditions in the amphibolite facies, and were likely deformed and intercalated within an accretionary prism formed during the Neoproterozoic to Cambrian closure of the Khomas Sea (Kukla and Stanistreet, 1991). The continuity of the Matchless amphibolite has been difficult to explain, but it may be by subduction of a mid-ocean ridge, where parts of the ridge were sliced off and accreted (Meneghini et al., 2014). Large metabasalt lenses, as in the photograph, have relatively high viscosity compared to surrounding metasediments. The scale of the phacoid in the photograph, and the existence of similar blocks of kilometers in largest dimension, implies that where topographic highs subducted, accretionary prisms may contain relatively high strength lenses at kilometers length scales. Cleavage wraps around these blocks, indicating ductile flow around phacoids, such that strain rate gradients between matrix and phacoids may elevate stress at the phacoid boundary. This boundary therefore becomes a likely zone of frictional failure at appropriate pressure–temperature conditions. As a consequence, the length scale of phacoids, from m to several km, may control the length scales of brittle failure in mixed-lithology deforming zones (Fagereng, 2011). *(Ake Fagereng)*

FIGURE 6.14 **The Sto. Domingo Anticline; the westermost termination of the South Pyrenean sole thrust.** Structural geologists use to simplify Nature to understand processes, however complex structures are common in fold and thrust belts; superposed, recumbent, plunging, conical folds, folded thrust, etc. This page is devoted to the San Marzal periclinal fold, a complex structure located in the External Sierras. The occurrence of Triassic evaporites acts as regional detachment level and it is responsible for a thin-skin deformation style. In this picture we can see a panoramic view of the Sto. Domingo anticline taken from the South (Puig Mone hill). The Internal Sierras (almost 3000m in altitude) are seen on the back. The Jaca molassic (Campodarbe Fm) and turbiditic (Hecho group) basin are located between both ranges. The Santo Domingo anticline is the more prominent structure in the western termination of the External Sierras front (Nichols, 1987; Millán et al., 1995). The Sto. Domingo anticline is a large-scale detachment fold that, due to the increasing amount of tectonic shortening to the East, is transformed into a thrust that can be tracked 40 additional kilometers. It is a 20km long fold (the westernmost 4 km are seen in this picture) striking WNW-NW and slightly oblique to the main Pyrenean trend. Apart from the western pericline termination, the rest of the structure seems to display a pseudo horizontal fold axis; although this may be partially conditioned by the outcropping conditions since most part of the hinge is eroded. *(Emilio L. Pueyo, Andrés Pocoví, Elisa M. Sánchez, Belén Oliva-Urcia)*

FIGURE 6.15 **San Marzal pericline; the ending of the Southwestern Pyrenean sole thrust.** In this picture we observe the magnificent termination of the External Sierras thrust front; the Sto. Domingo anticline and its lateral termination, the San Marzal pericline. The Guara Eocene limestones define the fold termination in the landscape while the Arguis marls produce a depression. 4500 m of fluvial deposits (Campodarbe Fm) cover the anticline core. Although San Marzal might seem a regular fold-closure in the geologic map, another significant feature from this panoramic view is the noteworthy westward plunge of the fold termination; the fold axis is sinking beneath our position. Serial balanced cross sections in the western part of the Sto. Domingo Anticline (Millán et al., 1995; Oliva-Urcia et al., 2012) characterize the wavelength (8-10 km) and amplitude (7 km) of the fold. The overall geometry may be fitted to a South verging cylindrical fold with pseudo-parallel and near-vertical limbs. However, the fold axis undergoes dramatic changes in shortening along-strike that are responsible for significant clockwise rotations of the northern limb. In summary, this complex geometry seems to better fit to a conical fold with an elliptical cross-section describing parallel near-vertical limbs (Millán et al., 1995) caused by the lateral gradient of shortening (Pueyo et al., 2003) together with the lateral disappearance of the detachment level to the West. All these facts have produced a pinning effect and the superb conical geometry of the fold. *(Emilio L. Pueyo, Andrés Pocoví, Elisa M. Sánchez, Belén Oliva-Urcia)*

FIGURE 6.16 **Active slope morphology operated by variably oriented fracture sets on the 1400-m high East wall of Monte Viso (3841 m).** Early winter snow makes the structures more shining in a view from the town of Saluzzo (Western Alps of Southern Piedmont, Italy). Prominent debris cones at the base of intersecting megafractures are being fed by rock falls and snow avalanches along ice-free couloirs; the intense deglaciation that affected this region in the last millennia, is still rapidly degrading the in-rock permafrost relictual on most of the slopes at altitude >3000 m. The front wall parallels N–S striking regional fractures and faults. *(Guido Gosso)*

FIGURE 6.17 **Early lineations crossed by later crenulation lineations in Paleoproterozoic metasedimentary rock, lower unit of Lesser Himalayan Crystallines, Kameng valley, western Arunachal Himalaya.** GPS location: N27°19′59.3″; E92°26′18.5″ (Yin et al., 2010; Bikramaditya Singh and Gururajan, 2011). *(Bikramaditya Singh)*

FIGURE 6.18 **Plumose structure or hackle plumes in quartzite at the contact between the Delhi Supergroup with the Mesoproterozoic basement gneisses, near Shrinagar ~15 km from Ajmer toward northeast.** Direction of divergence of the hackle lines connote direction of propagation of fractures. See Hobbs et al. (1976), Twiss and Moores (2007), and Davis et al. (2012) for detail of these structures. *(Eirin Kar)*

FIGURE 6.19 **A typical augen gneiss/granite mylonite at Funatsu Shear Zone within the Hida Belt of southwest Japan (N36°20′3.5′, E137°18′0′).** Foliation made by elongated aggregates of fine-grained biotite and quartz ribbons is clear on the nearly horizontal surface. Pale pinkish grains: K-feldspar, white grains: plagioclase, black strings: biotite, gray lenticular parts: quartz. Scale: pencil 14 cm. The Funatsu Shear Zone is a late Triassic dextral ductile shear zone, which develops in SE margin of the Hida Belt, consisting mainly of augen gneiss (deformed Hida older granites). Hida older granites date ~250 Ma and mylonitization happened in Late Triassic under upper greenschist to amphibolite facies (Takahashi et al., 2010). The Funatsu Shear Zone is regarded as the NE extension of the Cheongsan Shear Zone in the Korean Peninsula, which is regarded to be formed by collision between the north and the south China cratons (Takahashi et al., 2010; Ree et al., 2001). Therefore, the Funatsu Shear Zone is important to study late Paleozoic to Mesozoic tectonics of East Asia. *(Yutaka Takahashi)*

FIGURE 6.20 **Bedding–cleavage relation associated with regional folding.** The relation is found in beds dipping toward SE. The pen is parallel to the cleavage. Bedding dips more steeply than cleavage indicating that we are in the overturned limb of a larger-scale fold. See Till et al. (2008) and references therein for geology. Brooks Range, Alaska. N67°39.142/, W152°17.179/. Hunt Fork Shale, Late Devonian (Dillon et al., 1986). *(Victoria Pease)*

FIGURE 6.21 **Beautiful granodiorite mylonite in the coastal area of the Shirakami Mountains of north Japan (N40°27′47′, E139°56′40′).** These mylonites originate from the Shirakami-dake Granites, which are the members of the Abukuma Granites. K–Ar ages (biotite and hornblende) of the Shirakami-dake Granites are ~90 Ma (Fujimoto and Yamamoto, 2010). The mylonitization may have taken place ~90 Ma under upper greenschist to amphibolite facies condition. At this locality, granite mylonite is situated like a dyke in granodiorite mylonite, in which the foliations are parallel mutually. This occurrence indicates that the biotite granite had intruded into the hornblende-biotite granodiorite and these granite and granodiorite were mylonitized together (Takahashi, 2002). Scale bar: pencil 14 cm. "This mylonite zone in the Shirakami Mountains is a possible northern extension of the shear zone ~ Tanagura Tectonic Line, which is regarded as the NE extension of the Median Tectonic Line (Takahashi, 1999, 2002)". Therefore, granitic mylonites in the Shirakami Mountains are important to study Mesozoic tectonics of eastern margin of the Asian continent. *(Yutaka Takahashi)*

FIGURE 6.22 **Biotite rich granite with siltstone xenolith from the Tistung Formation.** U–Pb dating of zircons from the granite yields crystallization age at 480 ± 11 Ma (Gehrels et al., 2006). Location: Core of the Kathmandu klippe, Malekhu section, central Nepal Himalaya. *(Subodha Khanal)*

FIGURE 6.23 **Tension fracture surface with fine plumose structure with concentric ribs perpendicular to the plume.** The flaring fringe of hackles is at the edge of the joint surface. Each fringe consists of a series of alternating en-echelon cracks (also termed segments) and steps (also termed bridges) aligned along the parent joint (Frid et al., 2005). The fringe represents a direction of propagation of the fracture. This small exposure is located in a quarry near the village of Vapenna, Rychleby Mts, Czech Republic, Europe, Devonian crystalline limestones of Branna group (Novakova, 2008, 2010b). GPS coordinates: N50°16′35″, E017°07′49″. *(Lucie Novakova)*

FIGURE 6.24 **Rare structure named here "piggyback porphyroclasts?", in granite mylonite to protomylonite.** The structure shows two large porphyroclasts of K-feldspar occurring together in the central part of the photo and displaying strongly curved contacts due to crystal-plastic deformation. Notice the conjugate set of microfractures on the lower part of the contact zone inside the left porphyroclast. Subvertical curved contact between both porphyroclasts and linked conjugate microfractures suggest compression in the horizontal direction. Section oblique to the lineation and normal to the foliation. Width of view 206 mm. Location: Sierra de Velasco, La Rioja Province, NW Argentina. GPS point (WGS84): S 28°36′34.5″-W 67°10′41.1″ Rock type: Mylonite to protomylonite. Formation name: Ortogneiss Antinaco (TIPA Shear Zone) Age (relative): Post- Lower Ordovician / Pre- Carboniferous Facies/grade: medium grade *(Mariano A. Larrovere)*

FIGURE 6.25 **Spheroidally weathered Deccan basalt. Rounded fractures developed.** From the hill near Ananta building, IIT Bombay campus, Mumbai, India. *(Souvik Sen, Soumyajit Mukherjee)*

FIGURE 6.26 **Three nonparallel fractures intersect to develop a triangular zone in Deccan basalt, near Ananta building, IIT Bombay campus, Mumbai, India.** *(Souvik Sen, Soumyajit Mukherjee)*

FIGURE 6.27 **Remarkably rounded fracture plane, traced by placing pens, inside Deccan basalt.** This curved fracture plane is not a product of spheroidal weathering. Near Ananta building, IIT Bombay campus, Mumbai, India. *(Souvik Sen, Soumyajit Mukherjee)*

FIGURE 6.28 **Extension (tensile) joints in Archean granite gneisses near Perambalur, 50 km north of Tiruchirappalli, Tamil Nadu, India.** Notice their abutting relations. Three sets of joints: an E–W trending numbered 1, N–S trending numbered 2, and an E–W trending numbered 3. The joint set 1 seems oldest and 3 youngest. Extension direction is indicated perpendicular to the joint. Tensile joints decode extension directions. Pencil for scale, approximate length 10 cm. Location 11°13′3.53″N, 78°45′40.84″E. *(Achyuta Ayan Misra)*

FIGURE 6.29 **Basalt porphyry from a dike within Dharwar rocks.** Near Chakrapeta village, Kadapa district, Andhra Pradesh, India. Large phenocrysts of plagioclase feldspar are within fine grained matrix. The large phenocrysts in this photograph indicate a two-stage cooling. *(Atanu Mukherjee)*

FIGURE 6.30 **Axial planar cleavage in a small-scale fold.** Penetrative axial planar cleavage approximately parallel to the pen. Well developed in smaller-scale folds that mimic the regional direction of thrusting. See Till et al. (2008) and references therein for regional geology. Brooks Range, Alaska. N67°33.673/, W152°13.569/, Hunt Fork shale, Late Devonian (Dillon et al., 1986). *(Victoria Pease)*

FIGURE 6.31 **Bedding–cleavage relationship.** Well-developed S-asymmetry cleavage in mudstone layers interbedded with limestone (Jurassic). Esclangon, Pays Dignois, France. Left side: north direction. *(Amy Ellis)*

FIGURE 6.32 **Recrystallization of plagioclase in granodiorite.** Granodiorite of the Eagle Wash Intrusive Complex represents syntectonic Miocene (c. 19 Ma) magmatism followed by rapid uplift and cooling during a period of detachment faulting (Pease and Argent, 1999; Pease et al., 1999). Note low-temperature plasticity of the large plagioclase crystal in the central part of the image. Undulose extinction, deformation bands, and subgrain recrystallization along deformation bands suggest >450 °C temperature (Simpson, 1985; Brodie and Rutter, 1985). Sacramento Mountains, California, SW US. Coordinates: N3853884, E710016. *(Victoria Pease)*

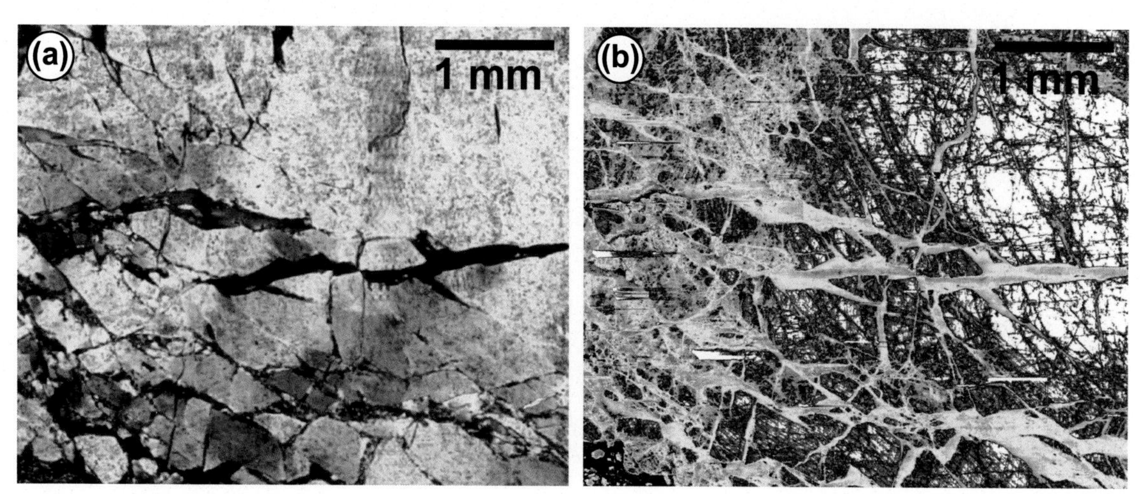

FIGURE 6.33 **Microstructural evidence for repeated cycles of coseismic damage and viscous flow at the frictional to viscous transition in an ancient strike-slip fault.** Large-displacement continental strike-slip faults are characterized by a core of ultrafine grained rocks surrounded by a "damage-zone" of fractured rocks that can extend laterally outward for hundreds of meters. A notable characteristic of these faults is the extensive occurrences of so-called "pulverized" rocks that exhibit intragranular cracking and fragmentation down to the micron scale. A fundamental question is how deep do these damage zones extend into the crust, and how do their physical characteristics change with depth (Mooney et al., 2007; Tullis et al., 2007)? The Paleozoic Norumbega Fault System in Maine, USA, is known to be one of the best ancient analogs for the San Andreas Fault currently exposed at Earth's surface (Ludman and West, 1999; Sibson and Toy, 2006). The Norumbega is exposed at a range of depths along its length, and in central Maine the mylonites along different fault strands preserve spectacular evidence for repeated coseismic brittle damage followed by viscous flow, representing the frictional to viscous transition. We conclude from our studies of these rocks that damage zones extend throughout the seismogenic zone, but are difficult to detect by geophysical or optical methods at depth owing to rapid microcrack healing and interseismic viscous flow. Some microstructural evidence is visible optically, but most of it can only be observed using cathodoluminescence (CL) and electron backscatter diffraction (EBSD) analysis. This figure pair from the Sandhill Corner shear zone (Price et al., 2012) illustrate the efficacy of CL imaging in extracting detailed microstructural history that is not visible optically. Optical (a) cross-polarized light and CL (b) images of a shattered plagioclase grain. Optically, the grain shows at least 3 generations of cross-cutting quartz-filled cracks, and patchy extinction. The CL image reveals remarkable complexity not visible optically. The dark patches are extensive tensile microcrack networks in plagioclase. These are overprinted by a dense network of quartz-filled cracks, the finest of which are not visible optically. Plagioclase fragments are as fine as 1 micron. *(Scott Johnson)*

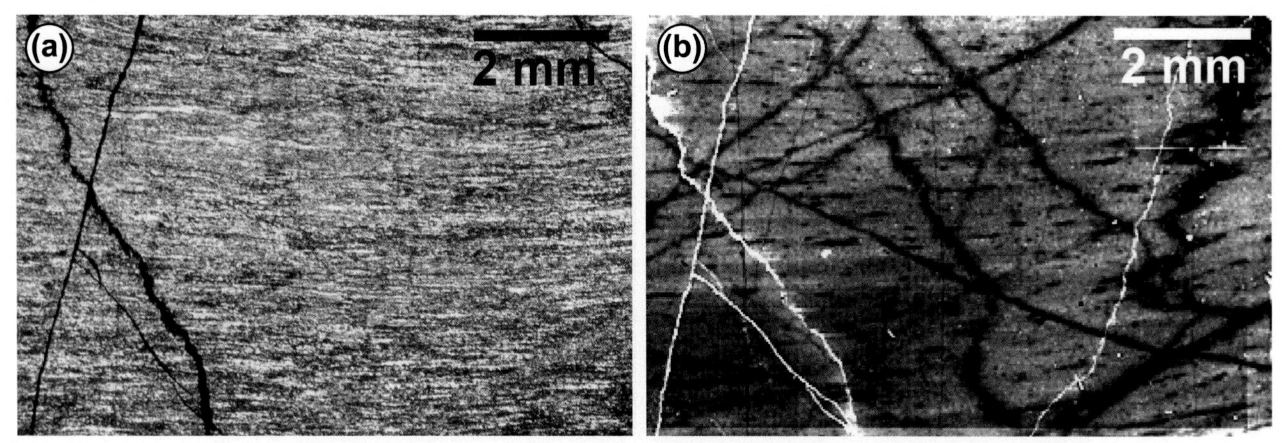

FIGURE 6.34 **Same locality as Figure 6.32. Optical (a) cross-polarized light and CL (b) images of a fine grained (<40 μm) recrystallized quartz ribbon.** Optically, the ribbon shows a typical recrystallized microstructure. The CL image shows a history of multiple overprinting fracture events preserved as dark bands in the quartz. Remarkably, the two earliest generations have been deformed and folded during viscous flow of the polycrystalline aggregate, suggesting at least 3 cycles of microcracking followed by viscous flow. All healed microcracks are lined with fluid inclusions. White lines in the CL image are open cracks, also visible optically. *(Scott Johnson)*

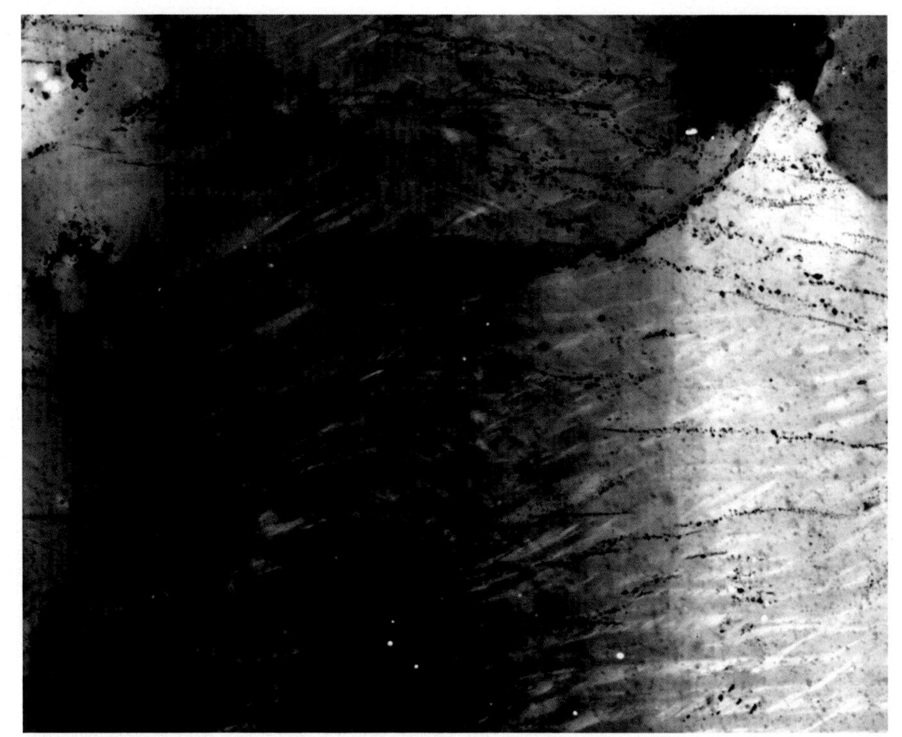

FIGURE 6.35 **Intracrystalline deformation microstructures in vein-quartz.** Location: Fosset (Belgium), vein-quartz in the Lower Devonian Anlier Formation of subgreenschist facies. Image width: 1 mm. Three sets of fine extinction bands in a single quartz crystal from an early orogenic quartz vein within quartzite deformed under subgreenschist conditions in the High-Ardenne slate belt (Belgium, France, Luxemburg, Germany), part of the Variscan orogenic belt. The optical axis is parallel to the thin section plane and is oriented top to bottom. The fine extinction bands orthogonal to the optical c-axis are broader (up to 10 μm) and more continuous than those inclined 40° to the left (up to 5 μm) and 40° to the right (up to 3 μm). C-axis misorientation between the fine extinction bands and the host crystal amounts up to 5°. Additionally, discontinuous fluid-inclusion trails run left to right through the image. Fine extinction bands in quartz have consistently been called "deformation lamellae" (Christie et al., 1964), for which tentatively, a maximum of two sets of fine extinction bands in a single crystal has been postulated resulting from a tectonic deformation (French and Koeberl, 2010; Hamers and Drury, 2011). In our tectonically deformed vein-quartz, though, we identify multiple cases of quartz single crystals with three sets of fine extinction bands, as shown in this image. Vernooij (2005) also produced three sets of FEBs with a similar morphology as in this image, in a quartz single crystal deformed during a compression experiment. We believe two sets of fine extinction bands are more common than suggested in literature. *(Tine Derez, Manuel Sintubin)*

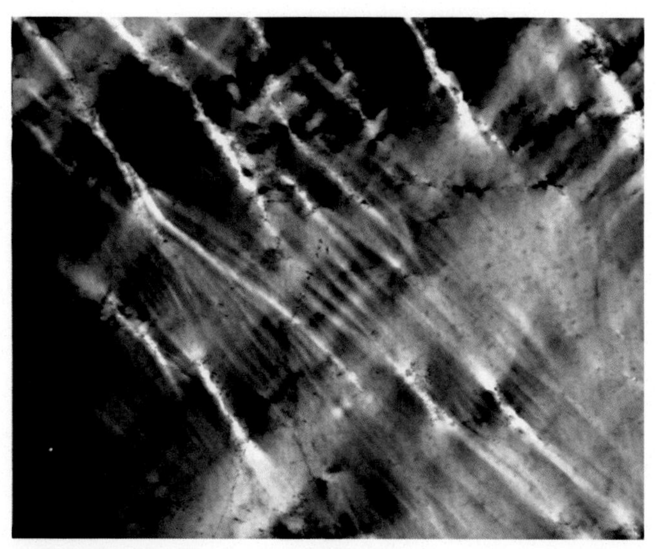

FIGURE 6.36 **Intracrystalline deformation microstructures in vein-quartz.** Location: Herbeumont (Belgium), vein-quartz from the Lower Devonian Anlier Formation of subgreenschist facies. Image width: 0.5 mm. Detail of a single quartz crystal in a late-orogenic quartz vein within metapelite deformed under subgreenschist conditions in the High-Ardenne slate belt (Belgium, France, Luxemburg, Germany), part of the Variscan orogenic belt. The optical axis is oriented top to bottom. The quartz single crystal contains three types of intracrystalline deformation microstructures. (1) Discontinuous localised extinction bands (LEBs) (~20 μm wide, 50° dip to the right, contain a high amount of decrepitated fluid-inclusions, c-axis misorientation with the host crystal up to 20°) continue into diverging fine (1- to 2-μm wide) extinction bands. (2) Orthogonal to the LEBs, ~50-μm wide extinction bands (3) run diagonally from top right to bottom left. LEBs mostly occur in conjugate, orthogonally arranged sets. Conjugate LEBs are present outside the frame of this image. Mostly LEBs terminate at the crystal boundary, or proceed in a narrow tail geometry within a zone of undulatory extinction bordered by fluid-inclusions. To our knowledge, LEBs have never been described to continue into diverging fine extinction bands before. Because these three deformation microstructures have been called a large variety of—often genetic—terms, we suggest to use new descriptive terminologies: *fine extinction bands*, *localised extinction bands*, and *wide extinction bands*, as suggested by Derez et al. (2015). *(Tine Derez, Manuel Sintubin)*

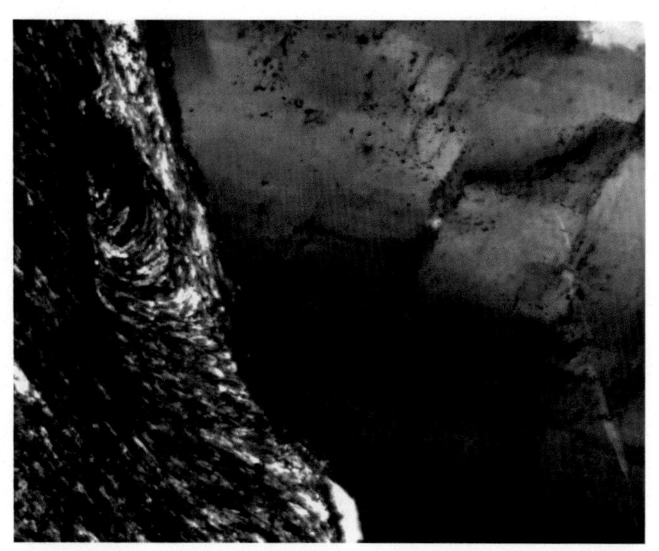

FIGURE 6.37 **Intracrystalline deformation microstructures in vein-quartz.** Location: Bertrix (Belgium), vein-quartz in the Lower Devonian Anlier Formation subgreenschist facies host-rock. Image width: 0.5 mm. Detail of a quartz single crystal from a late-orogenic quartz vein from the High-Ardenne slate belt (Belgium, France, Luxemburg, and Germany), part of the Variscan orogenic belt, in contact with the slaty host-rock. The quartz crystal, with the optical c-axis oriented approximately left to right, contains crosscutting fine extinction bands dipping 70° to the left and 20° to the right (~1 μm in width). A conjugate set of localised extinction bands (LEBs) with square to rectangular 20- to 200-μm wide subgrains, run top to bottom and top right to the center of the image. The rectangular subgrains are partly bound by en-echelon arranged fluid-inclusion planes dipping approximately 50° to the left. From diverse genetic interpretations suggested for LEBs, our observations favor the fracturation-shear model of Van Daalen et al. (1999). The current genetic interpretations need refinement, to explain the orientation of the fluid-inclusion planes, (1) dipping ~45° with respect to the LEBs orientation, (2) having the same orientation in both conjugate LEBs, and (3) being parallel to the maximum shortening direction as deduced by the method of Van Baelen (2010). Furthermore, the association between fine extinction bands and LEBs should be considered, as nearly all crystals containing LEBs consists of fine extinction bands (Derez et al., 2015). Visibility of fine extinction bands depends on the section. *(Tine Derez, Manuel Sintubin)*

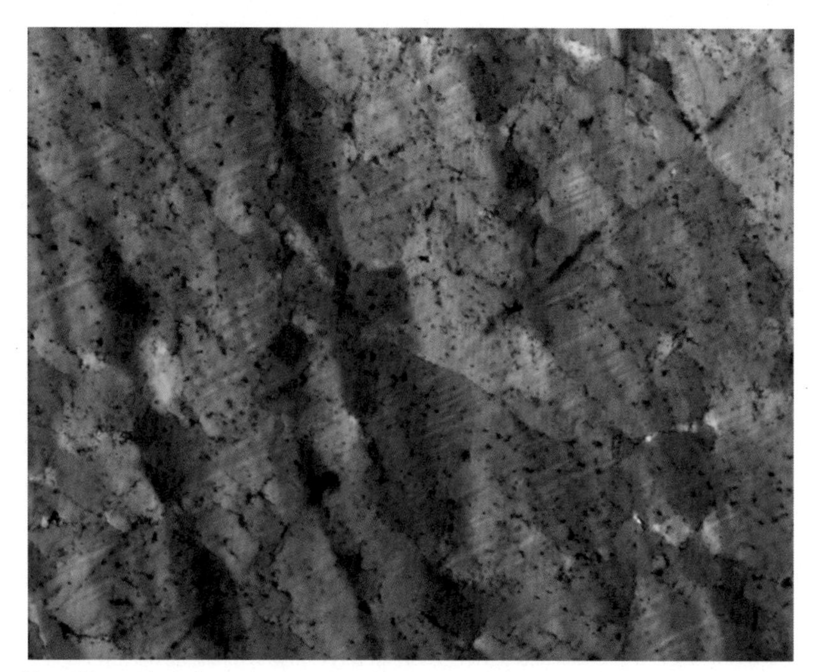

FIGURE 6.38 **Intracrystalline deformation microstructures in vein-quartz.** Location: Bertrix (Belgium), vein-quartz in Lower Devonian Anlier Formation subgreenschist facies host-rock. Image width: 0.5 mm. Detail of a quartz single crystal from a synorogenic vein within quartzite deformed under subgreenschist conditions in the High-Ardenne slate belt (Belgium, France, Luxemburg, Germany), part of the Variscan orogenic belt. Fine extinction bands (1 μm wide) with misorientation up to 5° and dipping ~15° to the left are orthogonal to wide extinction bands that are ~50-μm wide, and have a c-axis misorientation up to 20°, and dip 70° to the right. These parallel the optical axis. The wide extinction bands are partly bounded by fluid-inclusion planes perpendicular to their length, giving them a blocky character. Parallel to the wide extinction bands, localised extinction bands (LEBs) of rectangular subgrains, around 50 μm in width and c-axis misorientation up to 60°, can be observed, for example, in the upper left corner of the image. LEBs are very often observed to parallel wide extinction bands. In almost all crystals containing LEBs, fine extinction bands can almost always be observed, depending on the section. From this image it can be difficult to distinguish wide extinction bands with a blocky character from LEBs. Furthermore, in the vein-quartz of the High-Ardenne slate belt, adjoining wide extinction bands are observed to change into en-echelon LEBs. The same is also observed laterally along the wide extinction band orientation. Therefore, we believe LEBs exploit the orientation of previously formed wide extinction bands, as Gay (1974) suggested. *(Tine Derez, Manuel Sintubin)*

FIGURE 6.39 **Quartzite shows "grain boundary migration" (GBM) during grain coarsening (Barker, 1998).** This is relatively high temperature phenomenon in which grain boundaries are so mobile that they sweep through entire crystal to remove dislocations and produce subgrains. These grains are lobate and grain size varies as it is collected away from the Godhra granite. These grains are strain-free and show straight extinction (Passchier and Trouw, 2005). Ankalwa (23°09′38.56″N–73°37′27.02″E), Kadana Formation, Lunawada Group, Gujarat, India. *(Aditya Joshi, M.A.Limaye, Bhushan S. Deota)*

FIGURE 6.40 **Grain boundary area reduction (GBAR) in quartzite.** This is a high temperature phenomenon where quartz grains grow by accommodating adjoining smaller grains, produce straight boundaries and triple junctions (Passchier and Trouw, 2005). Such recrystallization is found close to the Godhra Granite. GBAR occurs during deformation and their effect is more obvious and may become dominant after deformation ceased (Bons and Urai, 1992). Chandsar (23°06′36.73″N-73°38′33.54″E), Kadana Formation, Lunawada Group, Gujarat, India. *(Aditya Joshi, M.A.Limaye, Bhushan S. Deota)*

FIGURE 6.41 **A "V" pull-apart microstructure of biotite in mylonite.** Margins of the two fragments are nonparallel. Based on the sense of rotation, it is interpreted as a "type 2a" "V" pull apart (Samanta et al., 2001). Within the "V," coarse- and fine-grained quartz occurs. See Mukherjee (2010, 2013) for more examples of V pull aparts. Natyal (23°05′39.51″N-73°42′06.63″E), Kadana Formation, Lunawada Group, Gujarat, India. *(Aditya Joshi, M.A.Limaye, Bhushan S. Deota)*

FIGURE 6.42 **Microfaulted plagioclase grain found within brecciated granite, equivalent to Closepet Granite, from a brittle shear zone.** The Microfaulting is subperpendicular to the twin lamellae of the feldspar grain. The plagioclase grain in this rock is surrounded by more or less elongated and recrystallized quartz grains. The feldspar deforms plastically a little before it fractures (brittle deformation), whereas the quartz flows and recrystallizes in a ductile manner (Vernon, 2004; Passchier and Trouw, 2005). Crossed polars. Width of photomicrogram: 2 mm. Sample from Gogi Village along Gogi-Kurlagere fault, Bhima Basin, India. *(Atanu Mukherjee)*

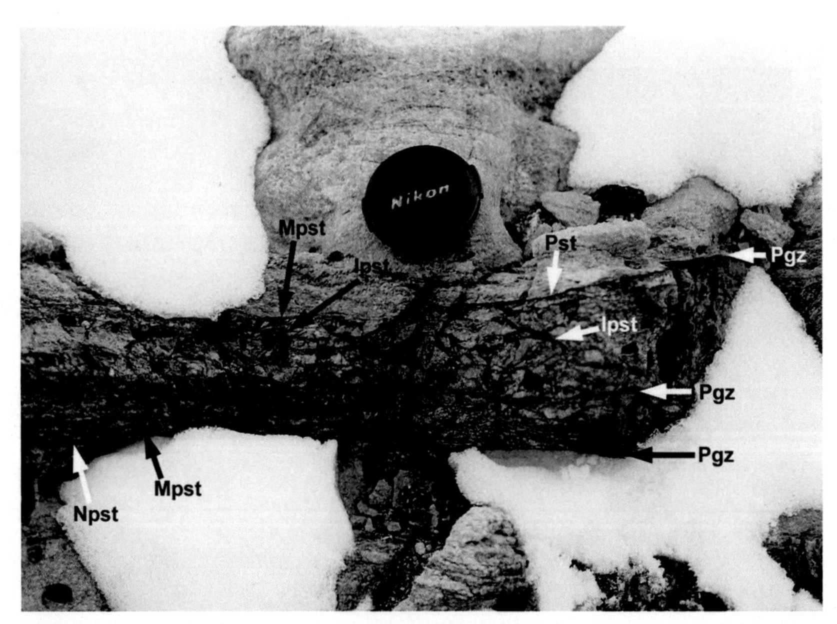

FIGURE 6.43 **Fault veins and injection veins of dark-colored pseudotachylyte derived from frictional melting and seismic faulting under granulite-facies conditions.** Three fault veins (Pst) of pseudotachylyte occur as pseudotachylyte-generating zones (Pgz) parallel to foliation of the surrounding garnet-bearing quartzofeldspathic gneiss and two pyroxene-garnet mylonite. Injection veins (Ipst) are oblique to the foliation of the host rocks and bear microlites of vermicular garnet, pyroxene, and plagioclase. Many fault veins of pseudotachylyte mylonitized (Mpst) intensely. The mylonites derived from the fault veins comprise granulite-facies mineral assemblages, including fine-grained orhopyroxene, clinopyroxene, and garnet in their matrix. However nonmylonitized pseudotachylytes (Npst) that were formed along the mylonitized fault veins often cut the mylonitic foliation. The new fault veins of pseudotachylytes also bear microlites of vermicular garnet, pyroxene, and plagioclase. These suggest seismic faulting and mylonitization took place repeatedly under granulite-facies conditions. Mylonitization under granulite-facies conditions occurred during Proterozoic time (Owada et al., 2001) or before 2550 Ma (Crowe et al., 2002). Location: 67°07′01″S, 50°19′07″E, Tonagh Island, Napier Complex in East Antarctica. *(Tsuyoshi Toyoshima, Yasuhito Osanai, Masaaki Owada, Toshiaki Tsunogae, Tomokazu Hokada)*

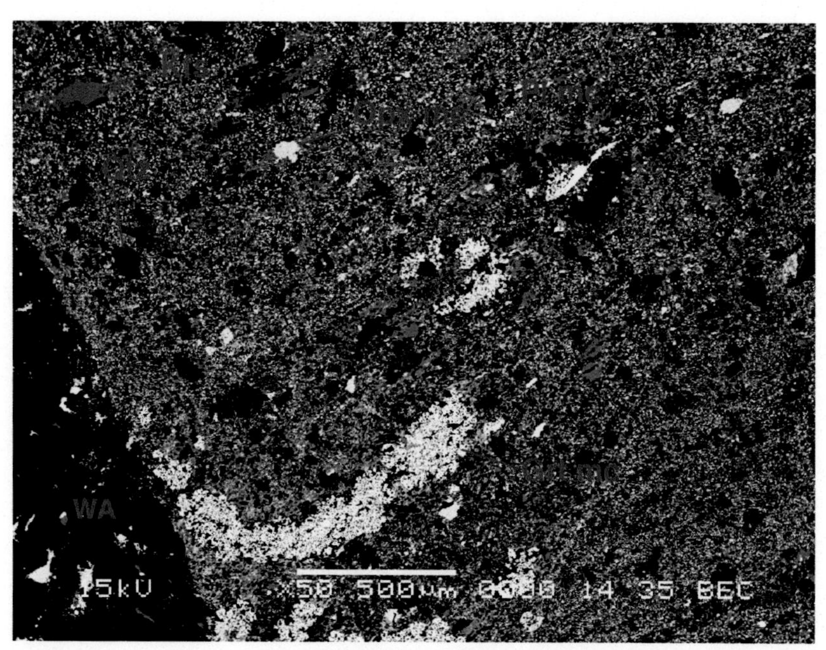

FIGURE 6.44 **Backscattered electron image of a part of injection vein of pseudotachylyte generated under granulite-facies conditions.** The injection vein of pseudotachylyte bears microlites of vermicular garnet (Grt mc), orthopyroxene (Opx mc), and plagioclase (Pl mc) but was very weakly mylonitized. However the fault veins of pseudotachylyte have been strongly mylonitized and changed to orthopyroxene-garnet mylonite under the granulite-facies conditions (Toyoshima et al., 1999). The granulite-facies mylonitization took place within ~2470–2550 Ma (Crowe et al., 2002) or during Proterozoic time (Owada et al., 2001). Kfs: K-feldspar fragment, Qtz: quartz fragment, WA: host metamorphic rock. Location: 67°06′04″S, 50°15′35″E, Tonagh Island, Napier Complex in East Antarctica. *(Tsuyoshi Toyoshima, Yasuhito Osanai, Masaaki Owada, Toshiaki Tsunogae, Tomokazu Hokada)*

FIGURE 6.45 **Amphibolite boudins folded and shortened by crustal shortening related to continental collision between East and West Gondwana.** Folded (Fb) and shortened (Sb) boudins of dark-colored amphibolite layers in gray-colored biotite-hornblende gneiss matrix vary in length within 5–20 cm. The spaces (necks) between the boudins are filled with white quartzofeldspathic veins. The boudins are upright folded with wavelengths up to ~30 cm. The vertical axial-plane foliations of the upright folds develop strongly on the right side of the photo. The boudins have originally pancake shapes with flattening parallel to compositional layering of the ambient gneiss, and resulted from the layer-normal shortening and thinning of crustal rocks between 640 and 600 Ma (Toyoshima et al., 1995). The folds are small-scale parasitic folds of a large-scale upright fold related to 600–560 Ma sinistral transpression and crustal shortening during the collision of parts of East and West Gondwana (Toyoshima et al., 2013). The folded and shortened boudins (Fb) suggest that the tectonic regime changed from layer-parallel extension to layer-parallel compression (Toyoshima et al., 2013). Osanai et al. (2013) presented SHRIMP U–PB ages for metamorphic rocks from the Sør Rondane Mountains, East Antarctica, and they recognized periods of ultrahigh-temperature metamorphism (pre-main metamorphic stage) within 750–700 Ma and granulite- to amphibolite-facies metamorphism during 640–600 Ma. Location: 72°09′40″S, 25°31′47″E, the southern part of Salen in the Sør Rondane Mountains, East Antarctica. *(Tsuyoshi Toyoshima, Masaaki Owada, Kazuyuki Shiraishi)*

FIGURE 6.46 **Mud cracks produced by desiccation in modern sediment are downward tapering, V-shaped fractures display crudely polygonal pattern in plan.** Formed in siliciclastic sediments along with imprints of raindrops. At Jhamarkotra mine, Ajmer, Rajasthan, India. *(Moloy Sarkar)*

FIGURE 6.47 **A very rare trapezoid shaped mud inclusion within sandstone.** Siwalik unit, near Mohand, Roorkee–Dehradun transect, Western Himalaya, Uttarakhand, India. Mukherjee (2012) recently reviewed trapezoid shaped objects in geology (also see Mukherjee 2013, 2014). *(Rajkumar Ghosh)*

FIGURE 6.48 **Filled sand cracks in Paleoproterozoic Gulcheru Formation is related to microbial mat destruction features (78°32′E and 14°20′ N).** Madyalabodu area, Gulcheru Formation. Cudappah basin, Andhra Pradesh, India. A network of intersecting cracks with varied geometry from polygonal/spindle shapes in mud-free Gulcheru sandstones suggests the presence of microbial mat contributed extra cohesion to the sand. They are formed on exposed surface due to desiccation of underwater in presence of microbial communities. These filling of cracks from above by other sands brought in subsequently by aqueous currents (Bottjer and Hegadorn, 2007). *(Sukanta Goswami)*

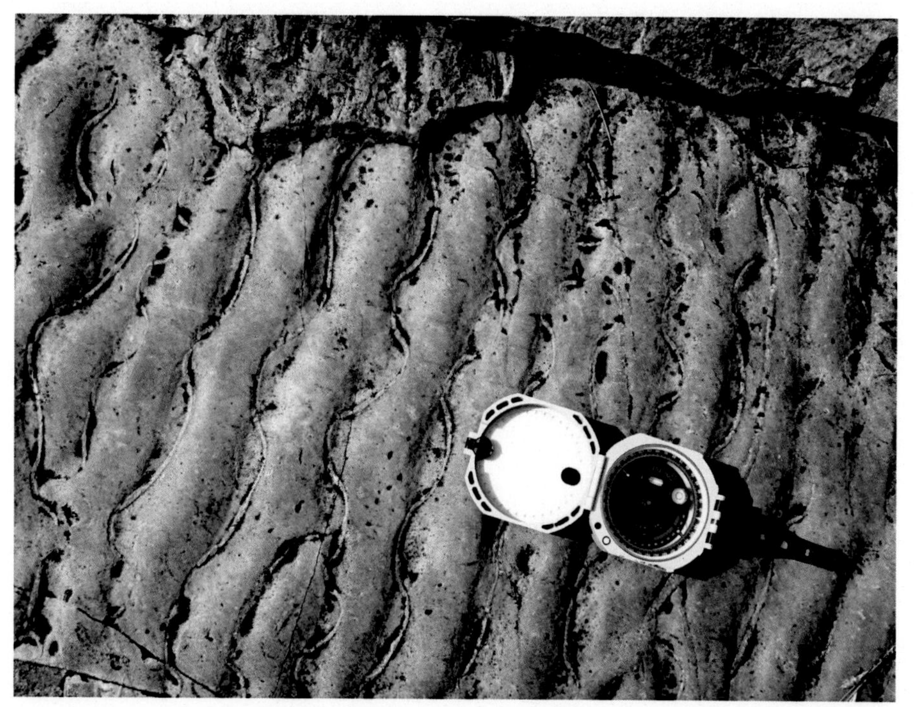

FIGURE 6.49 **This is a microbially induced sedimentary structure in Gulcheru Quartzite in SW margin of Cuddapah basin, 78°20′E and 14°18′ N, Giddankipalle area, Gulcheru Formation, Andhra Pradesh, India.** Within ripple troughs, complex crack patterns with sinuous forms occur. Those are called Manchuriophycus/Rhysonetron (Eriksson et al., 2007). Shrinkage cracks within rippled sandstone confined to ripple troughs may form if microbial mats grew selectively, or attain a greater thickness in the ripple troughs, where eroded cyanobacteria settle from currents under maximum and long lasting water supply. *(Sukanta Goswami)*

FIGURE 6.50 **Preservation of successive sets of ripples in sandy sediments with crests of each set being relatively sharp and unreworked.** Within Gulcheru Quartzite this successive ripple layers show switches in current directions. Absence of mud between successive rippled sandy beds is typical for palimpsest ripples. Nonamalgamation of two rippled sandstone beds can indicate the presence of biofilm or mat providing organically mediated sand stabilization. The preservation of underlying ripples despite deposition of overlying sandy rippled bed connotes presence of a microbial mat separating the two successive sand beds, protecting the earlier ripples from being reworked (Bottjer and Hegadorn, 2007). 78°30′E, 14°20′N, Madyalabodu area, Gulcheru Formation, Cudappah basin, Andhra Pradesh, India. *(Sukanta Goswami)*

FIGURE 6.51 **Ellipsoidal to bulbous chert nodules as structureless dense masses within carbonate rocks of Vempalle Formation, Motnutalapalle area, Cudappah basin, Andhra Pradesh, India.** They range a few centimeters up to 1 m. Occurrence of chert in Precambrian rocks indicating that they precipitated chemically from silica oversaturated water. Pore waters supersaturated with respect to silica and undersaturated with respect to calcium carbonate would favor the dissolution of calcium carbonate and the simultaneous precipitation of silica, especially with reduced pH (Knauth, 1979). Because alkaline marine surface waters are highly undersaturated with silica and nearly saturated with calcium carbonate, it is unlikely that chert nodules form from marine pore waters. Such conditions might occur under several sets of diagenesis involving meteoric pore waters enriched in silica from the dissolution of silicate minerals (Hefferan and O'Brien, 2010). *(Sukanta Goswami)*

FIGURE 6.52 **Characteristics of sedimentary dikes in lacustrine organic rich and poor laminated lime mudstone deposits of the Eocene Green River Formation, Washakie Basin, Wyoming.** Sedimentary dikes are up to 1.5 m long and several centimeters wide, filled with homogeneous lime mudstone or a mixture of fragmented material lacustrine sedimentary rocks and tuff material. In most cases the infill is silicified, and in cases a central conduit can be observed, filled with massive or laminated chert. If identifiable, the source of the cracks is a brecciated or massive silty lime mudstone layer, showing "syneresis"-like cracks. Often the dikes are represented by isolated blubs or horizontal sills in the laminated sediment, without a well-defined source interval. They show moderate sinuosity and branching. The dikes are more frequent and bigger in size at the base of a regional marker bed, and they can be traced for several tens of kilometers. Their morphological characteristics are similar to the ones found in several other startigraphic levels below, which suggests a common origin. Similar structures have been previously described as "dewatering structures" (e.g., Tänavsuu-Milkeviciene and Sarg, 2012), or sedimentary injection features (e.g., Hurst et al., 2011). Rhodes et al. (2007) interpreted these dikes as the result of desiccation. However the morphology and the infill of the cracks (with a central conduit) indicate dewatering, fluid migration, and sediment remobilization, which are related to tectonic activity at that time. Multiple filling indicates more than one fluidization event and remobilization of the sediments. *(Reproduced from Figure 22 of Törő et al. (2013). Balázs Törő, Brian R. Pratt, Jennifer J. Scott)*

FIGURE 6.53 **Archaeoseismology.** The instrumental record of seismic activity is less than 100 years old, while the recurrence of major earthquakes is measured on a centennial to millennial scale. Archaeoseismology aims to extend this record throughout the period characterized by artificial structures. Ancient buildings bearing particular features of damages provide evidence for past earthquakes. These allow to determine the parameters (date, intensity, epicenter, magnitude) of causative seismic events. Terms of structural geology are used in describing the damages produced by seismic loading. Examples of secondary, off-fault damages, produced by seismic shaking, are shown below on Figures 6.52 to 6.55. **V-shaped, extruded block bordered by conjugate faults.** A wedge-shaped block of masonry, c. $7 \times 7\,m^2$, shifted toward 240° azimuth by ~20 cm during the largest known earthquake in the Near East, in 1202 AD. Donjon of Al-Marqab citadel, Syria, built by the Hospitaller knights in the late twelfth century. Built of Roman concrete (rubble in lime), the external and internal walls are faced by dressed basalt blocks. Walls are up to 5 m thick and diameter of the tower is 20 m. Between floors—where the extruded wedge is—the building is compact. Al-Marqab, south of Baniyas on the Mediterranean coast, Syria (35°09′N, 35°56′E). See Kázmér and Major (2010) and Kázmér (2014) for detail. *(Miklós Kázmér)*

FIGURE 6.54 **Distributed left-lateral strike-slip deformation in the Roman theater of Baelo Claudia, Gibraltar Straits, Spain (36°05′N, 5°46′E).** There are conspicuous displacements visible on the wall of the staircase and between the blocks forming the arch in the foreground. Displacement is larger upwards (maximum 10 cm), indicating seismic shaking: the top part of the building oscillated with higher amplitude than the lower part, attached to the foundation. The town along the northern coast of Gibraltar Straits, flourished during the first to fourth century AD. After an earthquake of intensity IX-X MSK ~350–395 AD, the city was abandoned. Active faults nearby and a landslide within the area of the Roman town contributed to the damages. Scale: stairs on the left are 20–25 cm high. Read Silva et al. (2005), Kamh et al. (2008), Karakhanian et al. (2008), and Grützner et al. (2012) for detail. *(Miklós Kázmér)*

FIGURE 6.55 **Dropped keystones and adjacent blocks in the arches of the Colosseum, caused by earthquake shaking.** The feature is a miniature analogue of rifting: extension produced open fissures, normal faulting, subsidence, and tilting of blocks. The Flavian amphitheater in Rome, Italy, worldwide known as Colosseum (41°53′N, 12°29′E), was built between 70 and 76 AD. The external wall is composed of travertine blocks originally connected by iron pins and cramps embedded in lead, without mortar. Holes of uniform size between the blocks were made to steal the lead in times of scarce supply. Several earthquakes between 508 and 1703 AD affected Rome and the Colosseum. Heterogeneity of the soil beneath the foundations (lake sediments below the northern part as opposed to an abandoned bed of the river Tiberis below the southern part) yielded uneven vibration and settlement during earthquakes. This is why the southern perimeter wall of the Colosseum collapsed. The photo shows severely damaged arcades 23 and 24 between the collapsed and intact portions of the external wall on the eastern end of the amphitheater. These were closed by a brick wall during a restoration effort in 1805–1807, preserving the arches in their damaged form. Read Korjenkov and Mazor (2003), Kamai and Hatzor (2008), Marco (2008) and Pau and Vestroni (2008) for detail. *(Miklós Kázmér)*

FIGURE 6.56 **Rotated blocks within the seawall of the crusader castle, Tartous, coastal Syria (39°53′N, 35°52′E).** The fortress, built in the twelfth century by the knights of the Temple, suffered the greatest damage probably during the 1202AD earthquake. Blocks are c. 70cm high. The counterclockwise rotation of four blocks together (left block toward the viewer) was caused by E–W strong seismic motion. There is a right-lateral strike-slip fault on the northern side of the leftmost blocks. Vertical component of seismic waves rotated the blocks when the overburden was momentarily relieved, and seismic loading overcame friction. Read Korjenkov and Mazor (1999), Meghraoui et al. (2003), Similox-Tohon et al. (2006), and Rodríguez-Pascua et al. (2011) for detail. *(Miklós Kázmér)*

FIGURE 6.57 **Carnegie Rupes is a >250km long thrust fault related landform on the innermost planet, Mercury.** It crosscuts several large and small impact craters, the largest one, the Duccio crater near the center of the image, has ~110km diameter. As plate tectonics do not operate on Mercury, landforms like Carnegie Rupes form in response to global contraction caused by secular cooling of the planet's interior. Over 5000 such landforms have been identified on Mercury so far. The total shortening accommodated by these landforms corresponds to a decrease of Mercury's radius by as much as 7km. This image is a mosaic of 48 individual photographs taken by the Mercury Dual Imaging System (MDIS) aboard the MErcury Surface, Space ENvironment, GEochemistry, and Ranging (MESSENGER) spacecraft. The image is shown in equirectangular projection centered at 58.31°N, 53.07°W. (Image credit: NASA/Johns Hopkins University Applied Physics Laboratory/Carnegie Institution of Washington). *(Christian Klimczak, Paul K. Byrne)*

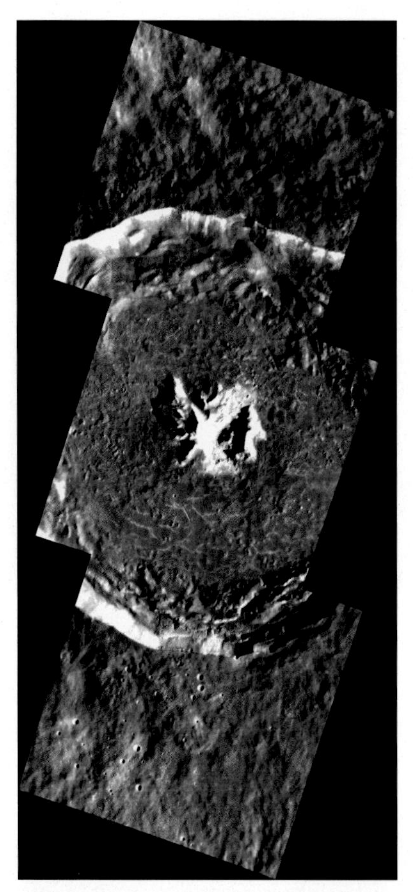

FIGURE 6.58 **The floor of Degas crater.** The floor of the well-preserved, 50-km-diameter Degas crater is covered with impact melt. As the impact melt sheet cooled, a generally hexagonal joint pattern developed. The central peak of Degas is made up of subsurface material uplifted by the impact. Over time, material from the tops of the central peaks has slid downslope. This exposed fresh material that appears bright in this image. The hummocky texture of the interior crater wall is due to landslides that developed after the crater formed. This mosaic, consisting of three Mercury Dual Imaging System (MDIS) photographs is shown in equirectangular projection centered at 36.99°N, 127.24°W. (Image credit: NASA/Johns Hopkins University Applied Physics Laboratory/Carnegie Institution of Washington). *(Christian Klimczak, Paul K. Byrne)*

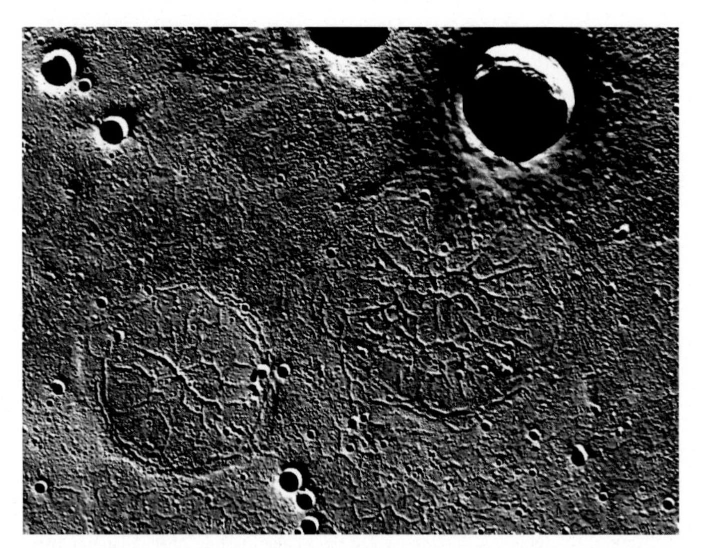

FIGURE 6.59 **The interior of the volcanically buried Goethe impact basin.** The Goethe impact basin is buried under a 1- to 2-km thick layer of lava. Two smaller craters near the basin center, 42 and 55 km in diameter, were also buried. The pooling and subsequent cooling of the thick lava sequence in these craters caused localized extension, and developed these large fractures. Circumferentially oriented fractures are superposed on rings of wrinkle ridges that likely demarcate the rims of the buried craters. Due to global contraction, Mercury is in a tectonic regime governed by large horizontal compressive stresses. Such a tectonic environment does not readily allow for extensional structures to form. Therefore, fractures such as those found in the volcanic units burying the Goethe basin can only form in areas that are protected from global contraction. This MDIS mosaic is shown in north polar stereographic projection; the center of the image is at 81.33°N and 52.32°W. (Image credit: NASA/Johns Hopkins University Applied Physics Laboratory/Carnegie Institution of Washington). *(Christian Klimczak, Paul K. Byrne)*

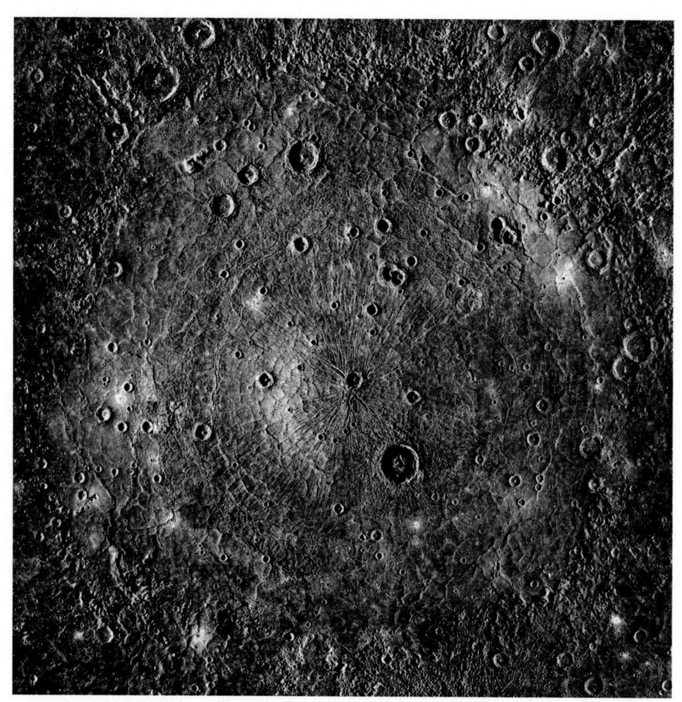

FIGURE 6.60 **The Caloris basin is the largest preserved impact structure on Mercury.** With an E–W diameter of 1640 km, the floor of the basin subtends ~40° of arc. The interior of the basin, which has been filled with voluminous, low-viscosity lavas, is replete with both contractional (wrinkle ridges) and extensional (graben) structures. The origin of these structures is not clear. The graben may have formed by thermal contraction of the interior lavas, similar to what is observed within Goethe basin, for example. Yet the pronounced radial orientations of the innermost graben are yet to be explained. The wrinkle ridges may be the result of subsidence of the Caloris interior lavas; however, in many places they remained active after the graben, and so could be due to the sustained global contraction of Mercury as its interior cooled. The image is in an orthographic projection centered at 31.5°N, 162.7°E. (Image credit: NASA/Johns Hopkins University Applied Physics Laboratory/Carnegie Institution of Washington). *(Christian Klimczak, Paul K. Byrne)*

REFERENCES

Barker, A.J., 1998. Introduction to Metamorphic Textures and Microstructures, second ed. Blackie, Glasgow. p. 264.

Bikramaditya Singh, R.K., Gururajan, N.S., 2011. Microstructures in quartz and feldspars of the Bomdila gneiss from western Arunachal Himalaya, India: implications for geotectonic evolution of the Bomdila mylonitic zone. Journal of Asian Earth Sciences 42, 1163–1178.

Bons, P.D., Urai, J.L., 1992. Syndeformational grain growth: microstructures kinetics. Journal of Structural Geology 14, 1101–1109.

Bottjer, D., Hegadorn, J.W., 2007. In: Schieber, J., Bose, P.K., Eriksson, P.G., Banerjee, S., Sarkar, S., Altermann, W., Catuneau, O. (Eds.), Atlas of Microbial Mat Features Preserved within the Siliciclastic Rock Record. Elsevier, pp. 53–71.

Brodie, K.H., Rutter, E.H., 1985. On the relationship between deformation and metamorphism, with special reference to the behavior of basic rocks. In: Thompson, A., Ruby, D. (Eds.), Metamorphic Reactions: Kinetics, Textures, and Deformation. Springer, pp. 138–179.

Brogi, A., Capezzuoli, E., 2009. Travertine deposition and faulting: the fault-related traver-tine fissure-ridge at Terme S. Giovanni, Rapolano Terme (Italy). International Journal of Earth Sciences 98, 931–947.

Chesterman, C.W., Kleinhampl, F.J., 1991. Travertine hot springs, Mono County, California. California Geology 44, 171–182.

Choukroune, P., Séguret, M., 1973. Carte structurale des Pyrénées, 1/500.000, Université de Montpellier–ELF. Aquitaine.

Christie, J.M., Griggs, D.T., Carter, B.J., 1964. Experimental evidence of basal slip in quartz. Journal of Geology 72, 734–756.

Crowe, W.A., Osanai, Y., Toyoshima, T., Owada, M., Tsunogae, T., Hokada, T., 2002. SHRIMP geochronology of a mylonite zone on Tonagh island: characterization of the last high-grade tectonothermal event in the Napier complex, East Antarctica. Polar Geoscience 15, 17–36.

Crossey, L.J., Karlstrom, K.E., 2012. Travertines and travertine springs in eastern Grand Canyon: what they tell us about groundwater, paleoclimate, and incision of Grand Canyon. Geological Society of America Special Papers 489, 131–143.

Davis, G.H., Reynolds, S.J., Kluth, C.F., 2012. Structural Geology of Rocks and Regions, third edition. Wiley, New York.

De Filippis, L., Billi, A., 2012. Morphotectonics of fissure ridge travertines from geothermal areas of Mammoth Hot Springs (Wyoming) and Bridgeport (California). Tectonophysics 548–549, 34–48.

De Filippis, L., Faccenna, C., Billi, A., Anzalone, E., Brilli, M., Özkul, M., Soligo, M., Tuccimei, P., Villa, I., 2012. Growth of fissure ridge travertines from geothermal springs of Denizli basin, western Turkey. Geological Society America Bulletin 124, 1629–1645.

De Filippis, L., Faccenna, C., Billi, A., Anzalone, E., Brilli, M., Soligo, M., Tuccimei, P., 2013. Plateau versus fissure ridge travertines from quaternary geothermal springs of Italy and Turkey: interactions and feedbacks between fluid discharge, paleoclimate, and tectonics. Earth-Science Reviews 123, 35–52.

Denyer, P., Alvarado, G.E., 2007. Mapa geológico de Costa Rica. Escala 1:400.000. Librería Francesa, San José, Costa Rica.

Derez, T., Pennock, G., Drury, M., Sintubin, M., 2015. Low-temperature intracrystalline deformation microstructures in quartz. Journal of Structural Geology 71, 3–23.

Dillon, J.T., Brosge, W.P., Dutro, J.T., 1986. Generalized Geologic Map of the Wiseman Quadrangle. US Geological Survey Open-File Report 86-219, Alaska.

Eriksson, P.G., Porada, H., Banerjee, S., Bouougri, E., Sarkar, S., Bumby, A.J., 2007. In: Schieber, J., Bose, P.K., Eriksson, P.G., Banerjee, S., Sarkar, S., Altermann, W., Catuneau, O. (Eds.), Atlas of Microbial Mat Features Preserved within the Siliciclastic Rock Record. Elsevier, pp. 76–105.

Faccenna, C., Soligo, M., Billi, A., De Filippis, L., Funiciello, R., Rossetti, C., Tuccimei, P., 2008. Late Pleistocene depositional cycles of the Lapis Tiburtinus travertine (Tivoli, central Italy): possible influence of climate and fault activity. Global and Planetary Change 63, 299–308.

Fagereng, A., 2011. Frequency-size distribution of competent lenses in a block-in-matrix melange: imposed length scales of brittle deformation? Journal of Geophysical Research 116, B05302.

French, B.M., Koeberl, C., 2010. The convincing identification of terrestrial meteorite impact structures: what works, what doesn't, and why. Earth-Science Reviews 98, 123–170.

Frid, V., Bahat, D., Rabinovich, A., 2005. Analysis of en echelon/hackle fringes and longitudinal splits in twist failed glass samples by means of fractography and electromagnetic radiation. Journal of Structural Geology 27, 145–159.

Fujimoto, Y., Yamamoto, M., 2010. On the granitoids in the Shirakami mountains and correlation to the Cretaceous to Paleogene granitoids distrtibuted in the Northeast Japan. Earth Science 64, 127–144. In Japanese with English abstract.

Gay, N.C., 1974. Modification of deformation lamellae during brittle-ductile deformation of quartzite. Geological Society of America Bulletin 85, 1237–1242.

Gehrels, G.E., DeCelles, P.G., Ojha, T.P., Upreti, B.N., 2006. Geologic and U-Th-Pb geochronologic evidence for early Paleozoic tectonism in the Kathmandu thrust sheet, central Nepal Himalaya. Geological Society of America Bulletin 118, 185–198.

Gratier, J.-P., Frery, E., Deschamps, P., Røyne, A., Renard, F., Dysthe, D., Ellouz- Zimmerman, N., Hamelin, B., 2012. How travertine veins grow from top to bottom and lift the rocks above them: the effect of crystallization force. Geology 40, 1015–1018.

Grützner, C., Reicherter, K., Hübscher, C., Silva, P.G., 2012. Active faulting and neotectonics in the Baelo Claudia area, Campo de Gibraltar (southern Spain). Tectonophysics 554–557, 127–142.

Hamers, M.F., Drury, M.R., 2011. Scanning electron microscope-cathodoluminescence (SEM-CL) imaging of planar deformation features and tectonic deformation lamellae in quartz. Meteorites & Planetary Science 46, 1814–1831.

Hancock, P.L., Chalmers, R.M.L., Altunel, E., Çakir, Z., 1999. Travitonics: using travertine in active fault studies. Journal of Structural Geology 21, 903–916.

Hefferan, K., O'Brien, J., 2010. Earth Materials. John Wiley & Sons Ltd, Wiley-Blackwell, UK.

Hobbs, B.E., Means, W.D., Williams, P.F., 1976. An Outline of Structural Geology. John Wiley & Sons.

Hurst, A., Scott, A., Vigorito, M., 2011. Physical characteristics of sand injectites. Earth-Science. Reviews 106, 215–246.

Knauth, L.P., 1979. A model for the origin of chert in limestone. Geology 7, 274–277.

Kamai, R., Hatzor, Y.H., 2008. Numerical analysis of block stone displacements in ancient masonry structures: a new method to estimate historic ground motions. International Journal for Numerical and Analytical Methods in Geomechanics 32, 1321–1340.

Kamh, G.M.E., Kallash, A., Azzam, R., 2008. Factors controlling building susceptibility to earthquakes: 14-year recordings of Islamic archaeological sites in old Cairo, Egypt: a case study. Environmental Geology 56, 269–279.

Karakhanian, A.S., Trifonov, V.G., Ivanova, T.P., Avagyan, A., Rukieh, M., Minini, H., Dodonov, A.E., Bachmanov, D.M., 2008. Seismic deformation in the St. Simeon monasteries (Qal'at Sim'an), northwestern Syria. Tectonophysics 453, 122–147.

Kázmér, M., Damages to ancient buildings from earthquakes. In: Beer M., Patelli E., Kouigioumtzoglou I., Au, I.S-K. (Eds.), Encyclopedia of Earthquake Engineering, Springer, Berlin, submitted for publication.

Kázmér, M., Major, B., 2010. Distinguishing damages of two earthquakes – archeoseismology of a Crusader castle (Al-Marqab citadel, Syria). In: Sintubin, M., Stewart, I., Niemi, T., Altunel, E. (Eds.), Ancient Earthquakes, vol. 471. Geological Society of America Special Paper, pp. 186–199.

Korjenkov, A.M., Mazor, E., 2003. Archeoseismology in Mamshit (Southern Israel): cracking a millennium-old code of earthquakes preserved in ancient ruins. Archäologischer Anzeiger 2003 (2), 51–82.

Korjenkov, A.M., Mazor, E., 1999. Seismogenic origin of the ancient Avdat ruins, Negev desert. Israel Natural Hazards 18, 193–226.

Krishnan, S., Ji, C., Komatitsch, D., Tromp, J., 2006. Case studies of damage to tall steel moment-frame buildings in southern California during large San Andreas earthquakes. Bulletin of the Seismological Society of America 96, 1523–1537.

Kukla, P.A., Stanistreet, I.G., 1991. Record of the Damaran Khomas Hochland accretionary prism in central Namibia: Refutation of an "Ensialic" origin of a Late Proterozoic orogenic belt. Geology 19, 473–476.

Ludman, A., West, Jr., D.P., (Eds.). 1999. The Norumbega fault system of the northern Appalachians. Geological Society of America Special Paper 331, 199.

Marco, S., 2008. Recognition of earthquake-related damage in archaeological sites: examples from the Dead Sea fault zone. Tectonophysics 453, 148–156.

Meghraoui, M., Gomez, F., Sbeinati, R., Van der Woerd, J., Mouty, M., Darkal, A.N., Radwan, Y., Layyous, I., Al Najjar, H., Darawcheh, R., Hijazi, F., Al-Ghazzi, R., Barazangi, M., 2003. Evidence for 830 years of seismic quiescence from paleoseismology, archaeoseismology and historical seismicity along the Dead Sea Fault in Syria. Earth and Planetary Science Letters 210, 35–52.

Meneghini, F., Kisters, A., Buick, I., Fagereng, A., 2014. Fingerprints of Late Neoproterozoic Ridge Subduction in the Pan-African. Geology, Damara belt, Namibia. Geology 42, 903–906.

Millán, H., Pocoví, A., Casas, A., 1995. El frente de cabalgamiento surpirenaico en el extremo occidental de las Sierras Exteriores: sistemas imbricados y pliegues de despegue. Revista de la Sociedad Geológica de España 8, 73–90.

Mooney, W., Beroza, G., Kind, R., 2007. Fault zones from top to bottom. In: Handy, M.R., Hirth, G., Hovius, N. (Eds.), Tectonic Faults - Agents of Change on a Dynamic Earth. The MIT Press, Cambridge, Mass., USA, pp. 2–46. Dahlem Workshop Report 95.

Mukherjee, S., 2010. V-pull apart structure in garnet in macro-scale. Journal of Structural Geology 32, 605.

Mukherjee, S., 2012. Simple shear is not so simple! Kinematics and shear senses in Newtonian viscous simple shear zones. Geological Magazine 149, 819–826.

Mukherjee, S., 2013. Deformation Microstructures in Rocks. Springer.

Mukherjee, S., 2014. Atlas of Shear Zone Structures in Meso-scale. Springer.

Muñoz, J.A., Beamud, E., Fernández, O., Arbués, P., Dinarès-Turell, J., Poblet, J., 2013. The Ainsa fold and thrust oblique zone of the central Pyrenees: kinematics of a curved contractional system from paleomagnetic and structural data. Tectonics 32, 1142–1175.

Nichols, G.J., 1987. The Structure and Stratigraphy of the Western External Sierras of the Pyrenees, Northern Spain. Geological Journal 22, 245–259.

Novakova, L., 2008. Main directions of the fractures in the limestone and granite quarries along the Sudetic Marginal Fault near Vápenná village, NE Bohemian Massif, Czech Republic. Acta Geodynamica et Geomaterialia 5, 49–55.

Novakova, L., Hájek, P., Šťastný, M., 2010. Determining the relative age of fault activity through analyses of gouge mineralogy and geochemistry: a case study from Vápenná (Rychlebské hory Mts.), Czech Republic. International Journal of Geosciences 1, 66–69.

Oliva-Urcia, B., Casas, A.M., Pueyo, E.L., Pocoví, A., 2012a. Structural and paleomagnetic evidence for non-rotational kinematics in the western termination of the External Sierras (southwestern central Pyrenees). Geologica Acta 10, 1–22.

Osanai, Y., Nogi, Y., Baba, S., Nakano, N., Adachi, T., Hokada, T., Toyoshima, T., Owada, M., 2013. Geologic evolution of the Sør Rondane Mountains, East Antarctica: collision tectonics proposed based on metamorphic processes and magnetic anomalies. Precambrian Research 234, 8–29.

Owada, M., Osanai, Y., Tsunogae, T., Hamamoto, T., Kagami, H., Toyoshima, T., Hokada, T., 2001. Sm-Nd garnet ages of retrograde garnet bearing granulites from Tonagh Island in the Napier Complex, East Antarctica: a preliminary study. Polar Geoscience 14, 75–87.

Pease, V., Argent, J., 1999. The northern Sacramento mountains, SW United States, Part I: structural profile through a crustal extensional detachment system. In: MacNiocaill, C., Ryan, P. (Eds.), Continental Tectonics, vol. 164. Geological Society of London Special Publication, pp. 179–198.

Pease, V., Foster, D., Wooden, J., O'Sullivan, P., Argent, J., Fanning, C., 1999. The northern Sacramento mountains, SW United States, Part II: exhumation history and detachment faulting. In: MacNiocaill, C., Ryan, P. (Eds.), Continental Tectonics, vol. 164. Geological Society of London Special Publication, pp. 199–237.

Pueyo, E.L., Parés, J.M., Millán, H., Pocoví, A., 2003a. Conical folds and apparent rotations in paleomagnetism (a case study in the Pyrenees). Tectonophysics 362, 345–366.

Passchier, C.W., Trouw, R.A.J., 2005. Microtectonics. Springer-Verlag, Berlin.

Pau, A., Vestroni, F., 2008. Vibration analysis and dynamic characterization of the Colosseum. Structural Control and Health Monitoring 15, 1105–1121.

Price, N.A., Johnson, S.E., Gerbi, C.C., West Jr., D.P., 2012. Identifying deformed pseudotachylyte and its influence on the strength and evolution of a crustal shear zone at the base of the seismogenic zone. Tectonophysics 518–521, 63–83.

Ree, J.H., Kwon, S.H., Park, Y., 2001. Pretectonic and posttectonic emplacements of the granitic rocks in the south central Okcheon belt, South Korea: Implications for the timing of strike-slip shearing and thrusting. Tectonics 20, 850–867.

Rhodes, M.K., Malone, D.H., Carroll, A.R., 2007. Sudden desiccation of Lake Gosiute at ~49 Ma: Downstream record of Heart Mountain faulting? The Mountain Geologist 44, 1–10.

Rodríguez-Pascua, M.A., Pérez-López, R., Silva, P.G., Giner-Robles, J.L., Garduño-Monroy, V.H., Reicherter, K., 2011. A comprehensive classification of earthquake archaeological effects (EAE) for archaeoseismology. Quaternary International 242, 20–30.

Samanta, S.K., Mandal, N., Chakraborty, C., 2001. Development of different types of pull-apart microstructures in mylonites: an experimental investigation. Journal of Structural Geology 24, 1345–1355.

Séguret, M., 1972. Etude tectonique des nappes et séries décollées de la partie centrale du versant sud des Pyrénées (PhD. thesis). Caractère synsédimentaire, rôle de la compression et de la gravité. University of Montpellier.

Silva, P.G., Borja, F., Zazo, C., Groy, J.L., Bardají, T., De Luque, L., Lario, J., Dabrio, C.J., 2005. Archaeoseismic record at the ancient Roman city of Baelo Claudia (Cádiz, south Spain). Tectonophysics 408, 129–146.

Sibson, R.H., Toy, V.G., 2006. The habit of fault-generated pseudotachylyte: Presence vs. absence of friction-melt. In: Earthquakes: Radiated Energy and the Physics of Faulting. Geophysical Monograph Series 170, 153–166.

Similox-Tohon, D., Sintubin, M., Muchez, P., Verhaert, G., Vanneste, K., Fernandez, M., Vandycke, S., Vanhaverbeke, H., Waelkens, M., 2006. The identification of an active fault by a multidisciplinary study at the archaeological site of Sagalassos (SW Turkey). Tectonophysics 420, 371–397. Erratum: 435, pp. 55–62.

Simpson, C., 1985. Deformation of granitic rocks across the brittle-ductile transition. Journal of Structural Geology 7, 503–511.

Stipp, M., Stunitz, H., Heilbronner, R., Schmid, S.M., 2002. The eastern Tonale fault zone: a 'natural laboratory' for crystal plastic deformation of quartz over a temperature range from 250 to 700 C. Journal of Structural Geology 24, 1861–1884.

Takahashi, Y., 1999. Reexamination on the northern extension of the Tanagura Tectonic Line, with special reference to the Nihonkoku-Miomote Mylonite Zone. Structural Geology 43, 69–78. In Japanese with English abstract.

Takahashi, Y., 2002. Granitic mylonites situated around the Shirakami Mountains. Northeast Japan. Earth Science 56, 215–216. In Japanese.

Takahashi, Y., Cho, D.L., Kee, W.S., 2010. Timing of mylonitization in the Funatsu Shear Zone within Hida Belt of southwest Japan: Implications for correlation with the shear zones around the Ogcheon Belt in the Korean Peninsula. Gondwana Research 17, 102–115.

Tänavsuu-Milkeviciene, K., Sarg, F.J., 2012. Evolution of an organic-rich lake basin – stratigraphy, climate and tectonics: Piceance Creek basin, Eocene Green River Formation. Sedimentology 59, 1735–1768.

Ternet, Y., Baudin, T., Laumonier, B., Barnolas, A., Gil-Peña, I., Martín-Alfageme, S., 2008. Mapa Geológico de los Pirineos a E. 1: 400.000. IGME–BRGM, Madrid-Orleans.

Till, A.B., Dumoulin, J.A., Harris, A.G., Moore, T.E., Bleick, H.A., Siwiec, B.R., 2008. Bedrock Geologic Map of the Southern Brooks Range, Alaska, and Accompanying Conodont Data. US Geological Survey Open-File Report 2008-1149. 54 pp.

Törő, B., Pratt, B.R., Renaut, R.W., 2013. Seismically induced soft-sediment deformation structures in the Eocene lacustrine Green River Formation (Wyoming, Utah, Colorado, USA) – a preliminary study. Poster abstract, Calgary. GeoConvention 2013: Integration Calgary. Poster abstract.

Toyoshima, T., Osanai, Y., Baba, S., Hokada, T., Nakano, N., Adachi, T., Otsubo, M., Ishikawa, M., Nogi, Y., 2013. Sinistral transpressional and extensional tectonics in Dronning Maud Land, East Antarctica, including the Sør Rondane Mountains. Precambrian Research 234, 30–46.

Toyoshima, T., Osanai, Y., Owada, M., Tsunogae, T., Hokada, T., Crowe, W.A., 1999. Deformation of ultrahigh-temperature metamorphic rocks from Tonagh Island in the Napier Complex, East Antarctica. Polar Geoscience 12, 29–48.

Toyoshima, T., Owada, M., Shiraishi, K., 1995. Structural evolution of metamorphic and intrusive rocks from the central part of the Sør Rondane Mountains, East Antarctica. Proceedings of the NIPR Symposium on Antarctic Geosciences 8, 75–97.

Treiman, A.H., 2008. Ancient groundwater flow in the Valles Marineris on Mars inferred from fault trace ridges. Nature Geoscience 1, 181–183.

Trepmann, C.A., Stockhert, B., Dorner, D., Moghadam, R.H., Kuster, M., Roller, K., 2007. Simulating coseismic deformation of quartz in the middle crust and fabric evolution during postseismic stress relaxation — An experimental study. Tectonophysics 442, 83–104.

Tullis, T.E., Bürgmann, R., Cocco, M., Hirth, G., King, G.C.P., Oncken, O., Otsuki, K., Rice, J.R., Rubin, A., Segall, P., Shapiro, S.A., Wibberley, C.A.J., 2007. Group report: rheology of fault rocks and their surroundings. In: Handy, M.R., Hirth, G., Hovius, N. (Eds.), Tectonic Faults - Agents of Change on a Dynamic Earth. Dahlem Workshop Report 95. The MIT Press, Cambridge, Mass, USA, pp. 183–204.

Twiss, R.J., Moores, E.M., 2007. Structural Geology. WH Freeman and Company. New York. pp. 43–45.

Uysal, I.T., Feng, Y., Zhao, J.X., Isik, V., Nuriel, P., Golding, S.D., 2009. Hydrothermal CO_2 degassing in seismically active zones during the Late Quaternary. Chemical Geology 265, 442–454.

Uysal, I.T., Feng, Y., Zhao, J.X., Altunel, E., Weatherley, D., Karabacak, V., Cengiz, O., Golding, S.D., Lawrence, M.G., Collerson, K.D., 2007. U-series dating and geochemical tracing of late Quaternary travertine in co-seismic fissures. Earth and Planetary Science Letters 257, 450–462.

Van Baelen, H., 2010. Dynamics of a progressive vein development during the late-orogenic mixed brittle-ductile destabilisation of a slate belt. Examples of the High-Ardenne slate belt (Herbeumont, Belgium). Aardkdundige Mededelingen 24, 221p.

Van Daalen, M., Heilbronner, R., Kunze, K., 1999. Orientation analysis of localized shear deformation in quartz fibres at the brittle–ductile transition. Tectonophysics 303, 83–107.

Van Noten, K., Claes, H., Soete, J., Foubert, A., Özkul, M., Swennen, R., 2013. Fracture networks and strike–slip deformation along reactivated normal faults in quaternary travertine deposits, Denizli Basin, western Turkey. Tectonophysics 588, 154–170.

Vernon, R.H., 2004. A Practical Guide to Rock Microstructure. Cambridge University Press.

Vernooij M.G.C. 2005. Dynamic Recrystallisation and Microfabric Development in Single Crystals of Quartz during Experimental Deformation (Unpublished PhD. thesis), Eidgenössische Technische Hochschule, Zürich.

Yin, A., Dubey, C.S., Kelty, T.K., Webb, A.A.G., Harrison, T.M., Chou, C.Y., Celerier, J., 2010. Geologic correlation of the Himalayan orogeny and Indian craton: Part 2: structural geology, geochronology, and tectonic evolution of the Eastern Himalaya. Geological Society of America Bulletin 122, 360–395.

Author Index

Note: Page numbers followed by "f" indicate figures.

A

Abe, S., 107, 119f
Acharyya, S.K., 4f
Adachi, T., 1f, 147f
Agard, P., 96f
Agnon, A., 16f
Ahuja, H., 61f
Al Najjar, H., 153f
Alberti, M., 30f
Al-Ghazzi, R., 153f
Alsop, G.I., 1
Altunel, E., 127f–128f
Alvarado, G.E., 126f
Andersen, T., 86f, 102f–103f
Andersen, T.B., 86f, 103f
Anderson, J.R., 27f
Andreani, M., 97f
Angelier, J., 93f–94f
Anzalone, E., 127f
Arbués, P., 129f
Argent, J., 92f, 141f
Ashwal, L., 103f
Austrheim, H., 86f, 102f–103f
Avagyan, A., 152f
Azzam, R., 152f

B

Baba, S., 1f, 147f
Bachmanov, D.M., 152f
Bahat, D., 136f
Banerjee, M., 61f
Banerjee, S., 149f
Barazangi, M., 153f
Bardají, T., 152f
Barker, A.J., 144f
Barker, S.L.L., 97f
Barnes, S., 107f
Barnolas, A., 5f, 129f–130f
Baronnet, A., 97f
Barrier, E., 93f–94f
Barrier, P., 43f–44f
Baudin, T., 5f, 129f–130f
Beamud, E., 129f
Belardi, G., 98f
Bell, T.H., 1, 15f
Ben M'Barek, M., 4f
Bernoulli, D., 113f
Beroza, G., 98f, 141f
Bhadra S., 49, 61f
Bhaskar Rao, Y.J., 29f
Bhattacharya, G., 82f
Bikramaditya Singh, R.K., 15f, 51f, 133f
Billi, A., 98f, 127f–128f
Biswal, T.K., 61f
Biswas, R., 49
Bleick, H.A., 135f, 140f
Bons, P.D., 1, 145f
Borja, F., 152f
Bornemann, O., 39f
Bose, M.K., 27f
Bose, N., 82f
Bottjer, D., 149f–150f
Boullier, A.M., 97f
Bouougri, E., 149f
Breton, J.-P., 119f
Brilli, M., 127f
Brodie, K.H., 141f
Brodzikowski, K., 21f
Brogi, A., 127f
Brosge, W.P., 135f, 140f
Buick, I., 131f
Bumby, A.J., 149f
Burchardt, S., 55f
Bürgmann, R., 98f, 141f
Burov, E., 96f
Byrne, T., 79

C

Caine, J.S., 79f
Çakir, Z., 127f
Calamita, F., 4f, 17f, 100f
Calvo, J.P., 16f
Cantarelli, V., 130f
Capezzuoli, E., 127f
Capponi, G., 97f
Carreras, J., 1
Carroll, A.R., 151f
Carter, B.J., 142f–143f
Casas, A., 132f
Casas, A.M., 130f, 132f
Cawood, P., 62f
Celarc, B., 103f
Celerier, J., 133f
Cengiz, O., 128f
Chakraborty, C., 1, 145f
Chalmers, R.M.L., 127f
Chesterman, C.W., 127f
Chetty, T.R.K., 9f, 27f, 29f, 41f
Childs, C., 89f
Cho, D.L., 63f, 134f
Chou, C.Y., 133f
Choudhuri, M., 2f, 82f
Choukroune, P., 130f
Christensen, N.I., 110f
Christie, J.M., 142f
Claes, H., 128f
Clark, C., 29f, 49
Cocco, M., 98f, 141f
Collerson, K.D., 128f
Collins, A.S., 27f, 29f
Compagnoni, R., 96f
Corrado, S., 130f
Cortesogno, L., 97f
Cowan, H.A., 102f
Cox, S.F., 97f
Crispini, L., 97f
Crossey, L.J., 128f
Crowe, W.A., 28f, 61f, 146f–147f
Csontos, L., 89f

D

Dabrio, C.J., 152f
Dabrowski, M., 49f
Dal Piaz, G.V., 96f
Darawcheh, R., 153f
Darkal, A.N., 153f
Das, R., 61f
Datta, S., 61f
Davis, G.H., 49, 119, 120f, 134f
Deb, S.K., 38f
De Capitani, C., 86f, 103f
DeCelles, P.G., 136f
De Filippis, L., 127f–128f
De Keijzer, M., 38f
De Luque, L., 152f
Derez, T., 143f
De Vicente, G., 16f
de Wit, M.J., 27f
Decandia, F.A., 30f, 100f
Deiana, G., 100f
Deng, H., 3f
Denyer, P., 126f
Deschamps, P., 128f
Deseta, N., 103f
DeWaele, B., 61f
Di Vincenzo, M., 4f
Diener, J.F.A., 107f
Dillon, J.T., 135f, 140f

Dinarès-Turell, J., 129f
Doblas, M., 79, 93f
Dobmeier, C., 61f
Dodonov, A.E., 152f
Dorner, D., 143f
Druguet, E., 1
Drüppel, K., 64f
Drury, M., 143f
Drury, M.R., 142f
Drury, S.A., 9f
Dubey, C.S., 133f
Dumoulin, J.A., 135f, 140f
Dutro, J.T., 135f, 140f
Dutta, A., 61f
Dyni, J.R., 37f
Dysthe, D., 128f

E

Eggins, S.M., 97f
Egli, D., 113f
Ellouz-Zimmerman, N., 128f
El-Wahed, M.A.A., 83f
Engelder, T., 63f
Engelmann, R., 64f
Eriksson, P.G., 149f
Esestime, P., 4f, 17f
Evans, J.P., 79f
Evans, L.A., 1
Exner, U., 49
Ez, V., 1

F

Faccenna, C., 127f–128f
Fagereng, A., 11f, 62f, 107f, 131f
Fanning, C., 92f, 141f
Federico, L., 97f
Feng, Y., 128f
Fernandéz, C., 56f, 67f–68f
Fernandez, M., 153f
Fernández, O., 129f
Fiori, A.P., 100f
Fitzsimons, I.C.W., 49
Fletcher, R.C., 107
Fodor, L., 89f, 101f
Fogarasi, A., 101f
Forster, C.B., 79f
Fossen, H., 1, 89f, 113f
Fossen, H.A., 119
Foster, D., 92f, 141f
Foubert, A., 128f
Frehner, M., 49
French, B.M., 142f
Frery, E., 128f
Frid, V., 136f
Fügenschuh, B., 113f
Fujimoto, Y., 59f, 135f
Funiciello, R., 128f

G

Gadhavi, M.S., 90f
Gagan, M.K., 97f
Galbiati, B., 97f
Garduño-Monroy, V.H., 153f
Garfunkel, Z., 100f
Gawlick, H.-J., 113f
Gay, N.C., 144f
Gazdova, R., 101f
Gehrels, G.E., 136f
Gerbi, C.C., 98f, 141f
Gerdes, A., 65f
Ghosh, J.G., 27f
Ghosh, S.K., 26f, 35f, 38f, 60f, 108f
Gil-Peña, I., 5f, 129f–130f
Giner-Robles, J.L., 153f
Glodny, J., 64f
Godin, L., 1, 2f
Golding, S.D., 128f
Gomez, F., 153f
Gómez-Gras, D., 16f
Gomez-Rivas, E., 1
Gopalakrishnan, K., 9f
Goscombe, B.D., 107, 113f
Gosso, G., 96f, 109f–110f, 133f
Gotowała, R., 21f
Graseman, B., 49f
Grasemann, B., 2f
Gratier, J.-P., 97f, 128f
Greiling, R.O., 64f–65f, 100f
Griera, A., 1
Griggs, D.T., 142f–143f
Grimmer, J.C., 64f–65f
Groppo, C., 96f
Groy, J.L., 152f
Grujic, D., 49
Grützner, C., 152f
Gueguen, E., 30f
Gupta, S., 49, 61f
Gururajan, N.S., 15f, 133f

H

Hada, S., 62f
Hájek, P., 136f
Hamamoto, T., 146f–147f
Hamelin, B., 128f
Hamers, M.F., 142f
Han, R., 99f
Hancock, P.L., 93f–95f, 127f
Hand, M., 27f, 107
Harayama, S., 63f
Harris, A.G., 135f, 140f
Harris, L.B., 1, 2f, 28f, 61f
Harrison, T.M., 133f
Hatzor, Y.H., 152f
Havíř, J., 101f
Hawkins, J.E., 37f
Hazra, S., 26f, 35f, 60f
Healy, D., 49
Hefferan, K., 150f
Hegadorn, J.W., 149f–150f
Heilbronner, R., 125, 143f
Hemmann, M., 39f
Hibbard, M.J., 13f
Hibsch, C., 43f–44f
Hijazi, F., 153f
Hilgers, C., 119f
Hillary, G.W.B., 107f
Hirt, A., 64f
Hirth, G., 92f, 98f, 141f
Hobbs, B.E., 14f, 134f
Hofmeister, A.M., 49
Hokada, T., 1f, 146f–147f
Holdsworth, R.E., 1
Holland, M., 119f–120f
Holt, R.W., 9f
Hoogerduijn Strating, E.H., 97f
Horie, K., 27f
Huang, H., 41f
Huber, M.I., 32f
Huber, R., 96f
Hübscher, C., 152f
Hudleston, P.J., 1
Hurst, A., 151f

I

Insley, M.W., 4f
Invernizzi, C., 130f
Ishikawa, M., 1f, 147f
Isik, V., 128f
Ito, M., 62f
Ivanova, T.P., 152f

J

Jaroszewski, W., 90f
Jena, S.K., 61f
Jercinovic, M.J., 14f
Ji, C., 125
John, G., 86f
Johnson, R.C., 37f
Johnson, S.E., 98f, 141f
Johnston, M.R., 102f
Jolivet, L., 86f, 96f, 103f
Jovanović, D., 113f

K

Kagami, H., 146f–147f
Kallash, A., 152f
Kamai, R., 152f
Kamh, G.M.E., 152f
Kamha, S.Z., 83f
Karabacak, V., 128f
Karakhanian, A.S., 152f
Karanth, R.V., 90f
Karkanas, P., 97f
Karlstrom, K.E., 128f
Karriya, Y., 63f
Katayama, I., 103f
Kázmér, M., 151f
Kee, W.S., 63f, 134f
Kelsey, D.E., 27f
Kelty, T.K., 133f
Kettermann, M., 107
Khan, D., 38f
Khan, K., 61f
Khanal, S., 22f
Kihm, N., 39f
Kind, R., 98f, 141f
King, G.C.P., 98f, 141f
Kirby, S.H., 21f
Kirschner, D., 119f
Kisters, A., 131f

Kleinhampl, F.J., 127f
Kluth, C.F., 49, 120f
Knauth, L.P., 150f
Koeberi, C., 142f
Koehn, D., 107
Kolinsky, P., 101f
Komatitsch, D., 125
Komazawa, M., 63f
Kontny, A., 64f
Korjenkov, A.M., 152f–153f
Koshimoto, S., 41f
Kövér, Sz., 89f, 101f
Koyi, H.A., 1, 2f–3f, 19f, 38f, 51f, 55f–56f, 59f, 82f, 112f
Krishnan, S., 125
Kronenberg, A.K., 21f
Krystyn, L., 113f
Kukla, P.A., 38f, 107f, 131f
Kunze, K., 143f
Kuster, M., 143f
Kwon, S., 41f
Kwon, S.H., 134f

L

Lan, L., 1, 68f
Landis, C.A., 62f
Lario, J., 152f
Laumonier, B., 5f, 129f–130f
Lawrence, M.G., 128f
Layyous, I., 153f
Leeming, P.M., 28f, 61f
Lein, R., 113f
Leonardo Evangelista Lagoeiro, 69f–70f, 76f, 110f
Lin, A., 21f
Lisker, F., 97f
Lister, G.S., 49
Llorens, M.-G., 1
Ludman, A., 98f
Luiz Sérgio Amarante Simões, 69f–70f, 76f, 110f
Lundquist, S.M., 110f

M

Ma, S., 99f
Maeder, X., 107
Mahadani, T., 1, 51f
Major, B., 151f
Malone, D.H., 151f
Mancktelow, N.S., 49, 57f
Mandal, N., 1, 38f, 145f
Mandal, S., 22f
Marco, S., 16f, 152f
Marshak, S., 63f
Martel, S., 2f
Martín-Alfageme, S., 5f, 129f–130f
Maruyama, S., 9f, 103f
Mazor, E., 152f–153f
McClay, K.R., 4f
Means, W.D., 14f, 134f
Meghraoui, M., 153f
Meneghini, F., 131f
Messiga, B., 97f
Miklavič , B., 50f
Millán, H., 132f
Minini, H., 152f
Misra, A.A., 79, 82f
Missoni, S., 113f
Mitra, S., 3f
Mizoguchi, K., 99f
Mo, Y., 6f
Moghadam, R.H., 143f
Mohanty, D.P., 41f
Montenat, C., 43f–44f
Mooney, W., 98f, 141f
Moore, T.E., 135f, 140f
Moores, E.M., 107, 134f
Mortimer, N., 11f
Mouty, M., 153f
Muchez, P., 119f–120f, 153f
Mukherjee, R., 1, 51f
Mukherjee, S., 1, 2f, 6f, 19f–20f, 22f, 38f, 49, 51f, 55f–56f, 59f, 65f, 71f, 79, 82f, 93f, 112f, 119, 145f
Mukhopadhyay, B., 27f
Mulchrone, K.F., 49
Muñoz, J.A., 129f

N

Nabelek, P.I., 49
Nagesh, P., 41f
Nakajima, Y., 103f
Nakano, N., 1f, 147f
Nakano, S., 63f
Nash, C.R., 28f, 61f
Navabpour, P., 93f–94f
Nelson, K.D., 11f
Nichols, G.J., 132f
Nicol, A., 89f
Nogi, Y., 1f, 147f
Novak, M., 120f
Novakova, L., 93f, 101f, 136f
Nuriel, P., 128f

O

O'Brien, J., 150f
Ojha, T.P., 136f
Okamoto, K., 103f
Oliva-Urcia, B., 132f
Oncken, O., 98f, 141f
Osanai, Y., 1f, 146f–147f
O'Sullivan, P., 92f, 141f
Otsubo, M., 1f, 147f
Otsuki, K., 98f, 141f
Ott d'Estevou, P., 43f–44f
Owada, M., 1f, 146f–147f
Özkul, M., 127f–128f

P

Pace, P., 17f
Palotai, M., 89f
Paltrinieri, W., 4f
Pamplona, J., 56f, 67f–68f
Panigrahi, M.K., 61f
Paola Ferreira Barbosa, 69f–70f, 76f, 110f
Parés, J.M., 132f
Park, C., 41f
Park, Y., 134f
Parkinson, C.D., 103f
Passchier, C., 2f, 13f, 20f
Passchier, C.W., 15f, 49, 55f, 60f, 65f, 79, 98f, 107, 113f, 119, 144f–146f
Pau, A., 152f
Paul, J., 39f
Payne, J.L., 27f
Pazdírková, J., 101f
Pease, V., 92f, 141f
Peinl, M., 39f
Pelorosso, M., 4f
Pennacchioni, G., 57f
Pennock, G., 143f
Pérez-López, R., 153f
Petit, J.P., 93f–95f, 97f
Phillips, D., 97f
Phillips, R.J., 6f
Piccardo, G., 97f
Pinkstone, J., 21f
Pizzi, A., 17f
Placer, L., 103f
Platt, J.P., 66f
Plavsa, D., 29f
Poblet, J., 129f
Pocoví, A., 132f
Podladchikov, Y.Y., 38f
Polino, R., 96f
Poljak, M., 81f
Porada, H., 149f
Pratt, B.R., 37f, 151f
Praveen, M.N., 41f
Price, N.A., 98f, 141f
Pueyo, E.L., 132f
Punekar, J.N., 1, 51f

R

Rabinovich, A., 136f
Radwan, Y., 153f
Raith, M., 61f
Ramsay, J.G., 1, 32f, 49, 96f
Rankin, L.R., 28f, 61f
Rattenbury, M.S., 102f
Ravna, E.J.K., 86f, 103f
Ree, J.H., 134f
Regenauer-Lieb, K., 49
Reicherter, K., 152f–153f
Remitti, F., 62f
Renard, F., 128f
Renaut, R.W., 37f, 151f
Renda, P., 30f
Reynolds, S.J., 49, 119, 120f, 134f
Rhodes, M.K., 151f
Rice, J.R., 98f, 141f
Ridley, J., 79f
Ridolfi, M., 17f
Robinson, D.M., 22f
Rodas, G.S., 126f
Rodrigues, B.C., 56f, 67f–68f
Rodríguez-Pascua, M.A., 16f, 153f
Roller, K., 143f
Rossetti, C., 128f

Rossetti, F., 97f–98f
Røyne, A., 128f
Rožič, B., 50f
Rubatto, D., 97f
Rubin, A., 98f, 141f
Rudenach, M.J., 15f
Rukieh, M., 152f
Rusciadelli, G., 17f
Rutter, E.H., 63f, 141f
Rykkelid, E., 89f

S
Saha, P., 4f
Saito, Y., 27f
Sajinkumar, K.S., 41f
Samanta, S.K., 1, 145f
Santosh, M., 9f, 27f, 29f, 41f
Sarg, F.J., 37f, 151f
Sarkar, S., 149f
Sasvári, Á., 89f
Sato, K., 9f
Satolli, S., 4f, 17f, 100f
Sawyer, E., 107f
Saxena, N., 119f–120f
Sbeinati, R., 153f
Scambelluri, M., 97f
Schefer, S., 113f
Scheltema, K.E., 14f
Schmalholz, S.M., 38f, 107
Schmeling, H., 55f
Schmid, D.W., 107
Schmid, S.M., 113f, 125
Scisciani, V., 4f, 17f
Scott, A., 151f
Segall, P., 98f, 141f
Séguret, M., 5f, 129f–130f
Sengupta, S., 26f, 35f, 38f, 60f, 108f
Shapiro, S.A., 98f, 141f
Sheth, H.C., 125f–126f
Shimamoto, T., 99f
Shimizu, I., 74f–75f
Shiraishi, K., 1f, 147f
Sibson, R.H., 11f, 62f, 98f, 102f, 141f
Silva, P.G., 152f–153f
Similox-Tohon, D., 153f
Simkhada, P., 22f
Simpson, C., 92f, 141f
Singanenjam, S., 41f
Sinha, N., 79
Sintubin, M., 143f, 153f
Siwiec, B.R., 135f, 140f
Sjöström, H., 55f
Smyth, J.R., 103f
Snoke, A.W., 49
Soete, J., 128f
Sokoutis, D., 107
Soligo, M., 127f–128f
Špaček, P., 101f
Speranza, F., 4f
Spray, J., 102f
Srivastava, D.C., 86f
Stanistreet, I.G., 107f, 131f
Šťastný, M., 136f
Stipp, M., 125
Stockhert, B., 143f
Strozyk, F., 38f
Stunitz, H., 125
Sudar, M., 113f
Suellen Olivia Cândida Pinto, 69f–70f, 76f, 110f
Švancara, J., 101f
Swanson, M., 102f
Swennen, R., 128f
Sýkorová, Z., 101f
Sztanó, O., 89f, 101f

T
Takahashi, Y., 59f, 63f, 134f–135f
Talbot, C., 55f
Talbot, C.J., 38f, 107
Tänavsuu-Milkeviciene, K., 37f, 151f
Tari, G., 101f
Tavarnelli, E., 4f, 17f, 30f, 100f
ten Grotenhuis, S.M., 49, 65f
Ternet, Y., 5f, 129f–130f
Theye, T., 97f
Till, A.B., 135f, 140f
Tjia, H.D., 79
Togo, T., 99f
Törő, B., 37f, 151f
Townsend, D.B., 102f
Toy, V.G., 98f, 141f
Toyoshima, T., 1f, 146f–147f
Tramutoli, M., 30f
Treagus, S., 68f
Treagus, S.H., 1
Treiman, A.H., 125
Trifonov, V.G., 152f
Tromp, J., 125
Trouw, R.A.J., 13f, 15f, 49, 60f, 65f, 79, 98f, 119, 144f–146f
Tsunogae, T., 27f, 29f, 41f, 146f–147f
Tuccimei, P., 127f–128f
Tullis, J., 92f
Tullis, T.E., 98f, 141f
Turtù, A., 100f
Twiss, R.J., 107, 134f

U
Upreti, B.N., 136f
Ura, J., 38f
Urai, J.L., 38f, 40f, 119f–120f, 145f
Ural, J., 107
Uysal, I.T., 128f

V
Valenta, J., 101f
Van Baelen, H., 143f
van Daalen, M., 143f
Van der Woerd, J., 153f
van Gent, H., 38f
Van Gent, H.W., 38f
Van Loon, A.J., 21f, 43f–44f
Van Noten, K., 128f
Vandycke, S., 153f
Vanhaverbeke, H., 153f
Vanneste, K., 153f
Vannucci, R., 97f
Vanossi, M., 97f
Venkatasivappa, V., 41f
Verhaert, G., 153f
Vernon, R.H., 51f, 146f
Vernooij, M.G.C., 142f
Vestroni, F., 152f
Vignaroli, G., 97f–98f
Vigorito, M., 151f
Villa, I., 127f
Virgo, S., 119f
Vissers, R.L.M., 66f, 97f
Vrabec, M., 103f

W
Waelkens, M., 153f
Walsh, J.J., 89f
Wang, C., 6f
Wang, S., 6f
Watterson, J., 89f
Weatherley, D., 128f
Webb, A.A.G., 133f
Wendzel, J., 39f
Wessel, Z.R., 79f
West, Jr. D.P., 98f, 141f
Whittington, A.G., 49
Wibberley, C.A.J., 98f, 141f
Wiersma, D.J., 60f
Willemse, J.M., 119f–120f
Williams, M.L., 14f
Williams, P.F., 14f, 134f
Williams-Straud, S.C., 39f
Woldřich, J.N., 101f
Wooden, J., 92f, 141f

X
Xiao, W.J., 29f

Y
Yakymchuk, C., 1, 2f
Yamamoto, M., 59f, 135f
Yamato, P., 96f
Yang, Q.Y., 41f
Yao, L., 99f
Yellappa, T., 29f, 41f
Yin, A., 4f, 133f
Yoshida, S., 74f
Yuen, D.A., 49

Z
Zachrisson, E., 100f
Zanella, F., 39f
Zartman, R.E., 27f
Zazo, C., 152f
Zeibig, S., 39f
Zhang, C., 3f
Zhang, Z.C., 41f
Zhao, J.X., 128f
Živčić, M., 81f
Zulauf, G., 39f
Zulauf, J., 39f
Zupančič, P., 81f

Subject Index

Note: Page numbers followed by "f" indicate figures.

A

Alkali feldspar, 19f
Antigorite fibers, 96f

B

Bedding-cleavage relationship, 140f
Boudins/mullions
  calc-schist
    boudinaged quartz veins in, 112f
    rectangular boudinaged quartz veins in, 112f
  chocolate tablet boudins, 111f
  composite boudins, 108f
  definition, 107
  fracture planes, boudinaged quartz vein within metagreywacke, 108f
  gray conglomerate overlying red mud horizon, Wine Strand, 116f
  low-grade metamorphic carbonate sequence, domino boudins within, 113f
  metaturbidites, boudinaged metabasalt layer enveloped by, 107f
  migmatite gneiss sequence, boudinaged amphibolite layer, 109f
  neoproterozoic quartzites, mullions and vertical columns, 114f–115f
  pinch and swell structures, 109f–110f
  precambrian orthoquartzite, Aravalli Supergroup, 116f
  quartz vein boudinaged, 111f
  ultramylonitized peridotite, microboudinage structure in, 110f
Brittle fault zones
  accretionary wedge, 92f
  active Hronov-Poříčí Fault Zone, main faults of, 101f
  along-dip segmented normal fault and fault-related fold, 89f
  amphibolite gneiss, fault plane of, 96f
  antiformal stack duplex, 84f
  antigorite fibers, 96f
  asbestos mineralization, structural control on, 97f–98f
  brittle tectonics, 93f
  centimeter-scale top-to-NW brittle reverse fault, 86f
  chloritoschist lens, 97f

Brittle fault zones (*Continued*)
  Coal Creek Fault zone/Central Front Range, Colorado, 79f
  coseismic damage and viscous flow, repeated cycles of, 98f
  Cretaceous Bhuj Formation, sand-shale sequence in the, 84f
  Cretaceous clastics, synlithification faults in, 101f
  curvilinear fault-bend folding, 83f
  definition, 79
  detachment faulting, 92f
  erosive tectonic tools, 95f
  experimental fault gouge, overlapped slip-zone structures in, 99f
  fault slickenside containing calcite slickenfibers, 93f
  fold-and-thrust tectonic belt, 90f
  gravitational effect, synsedimentary normal fault, 85f
  high-al omphacite, 103f
  Kaikoura, 102f
  mesoscopic conjugate set, 82f
  metamorphic veins, 97f
  NNE-SSW-trending Olevano-Antrodoco-Sibillini oblique thrust ramp, 100f
  polished fault surface, 95f
  pseudotachylytes (PSTs), 86f
    multiple generations of, 102f
  quartz and clay around quartz-clast, fine-grained matrix of, 100f
  quartzite pebble, N-dipping subvertical fault plane, 89f
  quartzite pebble, subvertical fault plane cuts across, 88f
  Raša fault, fault zone of, 81f
  SE-dipping subparallel curved fault planes, 88f
  sinistral strike-slip faulted pebble, 87f
  slickolites, 94f
  small-scale faults, uniform set of, 85f
  subvertical brittle shear Y-planes bound tangentially curved brittle P-planes, 82f
  tectonic grooves, striae of, 94f
  top-to-SW brittle sheared/reverse faulted sandstone, 90f
  white/red gouge, 81f
  Y- and P-brittle shear planes, fascinating development of, 87f
Brittle tectonics, 93f

C

Calc-silicate gneiss, 29f
Caloris basin, 155f
Chloritoschist lens outcropping, 97f
Crosscutting veins
  definition, 119
  dilatant faults, 120f
  gneissic foliations, quartz vein, 121f
  limestone
    high-pressure cell in, 119f
    semi-ductile shear zone in, 120f
  mylonitized gneiss, quartz-rich vein in, 121f
  quartz vein network, 123f
  thin-curved quartz vein, 122f

D

Dharwar rocks, Basalt porphyry from, 139f
Ductile shear zones, 1
  antigorite mylonite and asymmetric sheared composite olivine-pyroxene clasts, 73f
  around m-scale delta-type asymmetric-feldspar clast, 55f
  augen gneiss, post D2, syn-D4 ultramylonite shear zone in, 61f
  boudinaged quartz veins, 59f
  calc schist, type 1 flanking structure within, 55f
  chloritoid, syntectonic microscopic porphyroblast of, 70f
  classic shear band boudin, 56f
  definition, 49
  deformed amygdules, 74f
  deformed radiolarian fossils, 75f
  dilational jog, 62f
  domino style normal faults, 49f
  ductile sheared quartz vein, 51f
  extensional shear bands, 50f
  feldspar σ-type porphyroclast, 52f
  fine-grained recrystallized matrix, plagioclase feldspar porphyroclast in, 60f
  fold-boudin, 67f
    evolutive sequence of, 67f
  foliation parallel train of boudins, schistose quartzite, 52f
  fractured garnet porphyroblast, 72f
  Funatsu Shear Zone, augen gneiss/granite mylonite in, 63f
  garnet bearing schist, sigmoidal quartz lens in, 53f

Ductile shear zones (*Continued*)
granite mylonite, polished slab of, 59f
heterolithic sandstone-shale, cleavage refraction across, 63f
imbricate thrust structure in flysch, 103f
incipient localized dextral shear zones, 58f
initial fracture, typical paired shear zones to, 58f
K-feldspar, delta mantled porphyroclast of, 53f
kyanite-garnet gneiss, 66f
localized heterogeneous shear zone, 57f
mylonitic garnet-micaschists, asymmetric sheared granite pebble in, 65f
mylonitic micaschists, asymmetric quartz ribbon from, 73f
mylonitic peridotite cropping, 74f
mylonitic quartz-mica-feldspar-matrix, kinematically grown magnetite (mag) in, 64f
mylonitized micaschist, 65f
Num orthogneiss, biotite + kyanite-bearing micaschists from, 72f
olivine, microstructure of, 70f
paleoproterozoic S-type Lesser Himalayan Granitoids, plane polarized light of, 51f
parallelogram/sigmoid mud unit, 68f
plane polarized light, syntectonic microscopic garnet under, 71f
post-D3 sigmoidal quartz vein in amphibolite, Bastar Craton, Sambalpur, 50f
precursor heterogeneities, heterogeneous ductile shear zones on, 57f
quartzite veinlet, 54f
sheared quartzofeldspathic gneiss, asymmetric quartz porphyroclasts in, 61f
shear fabrics display, 75f
sigmoid inclusion pattern, syntectonically grown garnet with, 76f
soft-sediment, sigma clasts/S-C structures in, 66f
spiral, early foliations, sheared microscopic garnet with, 71f
stacked-folds-boudins, 68f
strike slip fault system, duplex structure in, 60f
top-to-right/west sheared intercalations, 56f
top-to-SW sheared sigmoid quartz veins, 62f
trapizoidal boudinaged quartz vein, 54f
ultramylonitic peridotite, delta-like structure in, 69f
ultramylonitic shear zone, titanite porphyroclast in, 69f
winged δ-microstructure of olivine, 76f
Y-planes in mafic schist, Sigmoid brittle P-planes bound by, 64f

E

Erosive tectonic tools, 95f

F

Folds
alternate layers of ferrugineous and quartzose materials, folded banded iron formation of, 11f
banded gneiss, small-scale folds in, 22f
Bastar Craton-Rengali Province boundary shear zone, rotation of F2 fold hinges, 28f
boudinaged sheath fold, 26f
Bowmore crenulation cleavage, 14f
calcareous matrix, quartzofeldspathic layers interlayered in, 35f
calc-silicate, doubly plunging round hinge isoclinals fold in, 27f
calc-silicate gneiss, 29f
chevron-like fold in amphibolite, dismantle block, 17f
continental collision, upright folds and folded boudins resulting from, 1f
curved axial plane, noncylindrical fold with, 34f
curved hinge line, plane noncylindrical fold with, 35f
Cuspate folds of quartz vein, Udaipur, Rajasthan, 8f
definition, 1
dolomitic host rock, tightly folded quartz vein in, 20f
dolomitic rock, folded quartz vein in, 30f
fault-bend-fold, 2f
ferrogenous sandstone, plunging fold in, 24f
folded anhydrite layers, 38f
folded layer of chert, 25f
folded migmatic rock, 9f
folded quartz vein, Bhedaghat, Jabalpur, 8f
folded quartz vein within meta-greywacke, 36f
fractured anhydrite, red salt in, 39f
garnetiferous banded gneiss, crenulation cleavage in, 22f
Gravifossum, 44f
Gulcheru Quartzite, 6f
Higher Himalayan gneisses, 20f
high-grade quartzite rock, type 3 interference pattern/hook-shaped pattern in, 42f
horizontal E-W fold plunges, complex folding with, 41f
hornblende gneiss, recumbent fold plunging toward E in, 24f
interlayered sandstones and mudstones, 11f
intrafolial folds, 23f
isoclinally folded chert layers, 26f
Karakoram fault, Ayishan detachment lies between, 6f
khaki green shale, Jitpur, 7f
kink folded intercalated layers of shales and limestones, 7f
lacustrine organic rich dolomitic lime mudstone/oil-shale deposits
disharmonic folding in, 37f, 43f
disturbed layer in, 16f
large-scale folded banded magnetite quartzite and magnetite garnet biotite schist intercalation, 10f
Los Cabos Series quartzites, similar folds within, 16f
lower greenschist facies quartzite rock, two-generation folding in, 42f
mafic mineral-rich gneiss, feldspathic layer in, 44f
Mesoproterozoic garnetiferous mica schist, cross-nichol of, 15f
mesoscale, recumbent fold in, 5f
mesoscopic disharmonic folding, 30f
metamorphosed granitic host rock, amphibole-rich layer in, 21f
metasedimentary rock, stretched quartz boudins in, 19f
meta-siltstone, 3f
meta-siltstone, disjunctive spaced cleavage in, 14f
mica gneiss, microfolded cordierite in, 13f
mica schist, folded quartzite intercalated with, 18f
Middle Siwalik, micaceous dafla sandstone of, 21f
morphology and genesis of, 1
neoproterozoic marble, conjugate folds in, 36f
neutral folding, 28f
nonuniform geometry, ptygmatic folds of, 33f
originally rectangular intraformational mudstone rip-up clast set, 37f
out-of-syncline fold accommodation fault, 3f
overprinting crenulation cleavages, 15f
parasitic fold, 23f
planar limbs and sharp/pointed hinge, chevron fold with, 10f
potassium salt, structures of, 40f
prominent axial depression, doubly plunging fold with, 33f
quartz vein, fractured ptygmatic fold of, 40f
rock salt, anhydrite diapir in, 39f
sheath fold, 25f
Siang fold-and-thrust belt, eocene limestone-shale sequences, 4f
subvertical (100/78S) ferruginous quartzite layer, elongated dome and basin structures on, 38f
thrust fault and related footwall syncline, Mt Prena, 4f
Tibetan Himalaya, folded strata in, 5f
tight isoclinal fold
intensely deformed lower limb of, 9f
limbs and hinge regions, 18f
tourmaline-bearing garnet-mica schist, angular asymmetric folds in, 13f
tri-shear fault-propagation folding, 2f
type III fold interference, 34f
type 2 interference pattern, 41f
type 3 interference pattern of, 32f
undeformed massive sandstone turbidite beds, rhythmic occurrence of, 12f
Variscan basement, fold interference pattern in, 32f
Fractures
developing triangular zone, three nonparallel fractures intersect to, 138f
fine plumose structure, tension fracture surface with, 136f
fracture planes, boudinaged quartz vein within metagreywacke, 108f
fractured anhydrite, red salt in, 39f
fractured garnet porphyroblast, 72f

Fractures (*Continued*)
initial fracture, typical paired shear zones to, 58f
placing pens/inside Deccan basalt, rounded fracture plane, 138f
quartz vein, fractured ptygmatic fold of, 40f

G

Grain boundary area reduction (GBAR), 145f
Grain boundary migration, 125
Grain-size reduced scapolite, 19f
Granodiorite of the Eagle Wash Intrusive Complex, 92f
Gran Sasso thrust system, 4f
Granulite-to amphibolite-facies metamorphism, 1f

H

Highly foliated garnetiferous gabbroic rocks, 9f

I

Intrafolial folds, 23f

K

Kumbhalgarh Formation, 8f, 36f

L

Lameta Ghat, 111f
Light-colored biotite-hornblende gneiss, 1f

M

Mesoscopic conjugate set, 82f

P

Planar subvertical axial plane, 10f

S

Shear zones
active slope morphology, 1400-m high East wall of Monte Viso, 133f
Akköy fissure-ridge, banded travertine across, 128f
amphibolite boudins folded and shortened by crustal shortening, 147f
ancient strike-slip fault, frictional to viscous transition in, 141f
augen gneiss, post D2, syn-D4 ultramylonite shear zone in, 61f
Bagaces Formation ignimbrites, curved columnar joints, 126f
Bastar Craton-Rengali Province boundary shear zone, rotation of F2 fold hinges, 28f
bedding-cleavage relationship, 140f
biotite in mylonite, "V" pull-apart microstructure of, 145f
biotite rich granite with siltstone xenolith, Tistung Formation, 136f
bulbous chert nodules as structureless dense masses, carbonate rocks of Vempalle Formation, 150f
caloris basin, 155f
coastal area of the Shirakami Mountains, granodiorite mylonite in, 135f
columnar joints/colonnade and entablature, 125f
dark-colored pseudotachylyte, fault veins and injection veins of, 146f
Denizli extensional basin, Kamara active fissure-ridge in, 127f
developing triangular zone, three nonparallel fractures intersect to, 138f
Dharwar rocks, Basalt porphyry from, 139f
dropped keystones and adjacent blocks, 152f
extension (tensile) joints in Archean granite gneisses, Perambalur, 139f
external sierras front, magnificent termination of, 132f
fine plumose structure, tension fracture surface with, 136f
floor of Degas crater, 154f
fold and thrust belts, 132f
fold-and-thrust system, 129f
Funatsu Shear Zone, augen gneiss/granite mylonite at, 63f, 134f
geothermal travertine area, Bridgeport, California, 127f
grain boundary area reduction (GBAR), 145f
grain coarsening, grain boundary migration (GBM) during, 144f
Gulcheru Quartzite, Cuddapah basin, 149f
incipient localized dextral shear zones, 58f
initial fracture, typical paired shear zones to, 58f
injection vein, backscattered electron image of, 147f
localized heterogeneous shear zone, 57f
Long Ridge, southeast flank of, 128f
matchless amphibolite, Kuiseb Canyon, Namibia, 131f
microfaulted plagioclase grain, 146f
mud cracks produced by desiccation in modern sediment, 148f
optical cross-polarized light and cathodoluminescence (CL), 142f
Paleoproterozoic Gulcheru Formation, filled sand cracks in, 149f
Paleoproterozoic metasedimentary rock, crenulation lineations in, 133f
Palghat Cauvery shear zone, 29f
piggyback porphyroclasts, 137f
placing pens/inside Deccan basalt, rounded fracture plane, 138f
plagioclase in granodiorite, recrystallization of, 141f
precursor heterogeneities, heterogeneous ductile shear zones on, 57f
quartzite, plumose structure/hackle plumes in, 134f
regional folding, bedding-cleavage relation associated with, 135f
Roman theater, distributed left-lateral strike-slip deformation in, 152f
sandstones, cleavage planes in, 130f
sandy sediments, ripples in, 150f
seawall of crusader castle, Tartous, 153f
sedimentary dikes in lacustrine, 151f
seismic activity, instrumental record of, 151f
semi-ductile shear zone in, 120f
shear zones, 1
single basaltic lava flow, distinct colonnade and entablature tiers, 126f
small-scale fold, Axial planar cleavage in, 140f
spheroidally weathered Deccan basalt, 137f
thrust fault related landform on innermost planet, Mercury, 153f
thrust sheet stacking, 130f
ultramylonitic shear zone, titanite porphyroclast in, 69f
vein-quartz, intracrystalline deformation microstructures in, 142f–144f
volcanically buried Goethe impact basin, 154f
within sandstone, trapezoid shaped mud inclusion, 148f
Sigmoid-shaped vein, 51f

T

Thrust sheet stacking, 130f

V

Volcanically buried Goethe impact basin, 154f

Printed in the United States
By Bookmasters